Manufacturing Change

Warwick Studies in Industrial Relations
General Editors: G.S. Bain, R. Hyman and K. Sisson

Also available in this series

Manufacturing Change

Industrial Relations and Restructuring

Edited by
Stephanie Tailby and Colin Whitston

Basil Blackwell

Copyright © Basil Blackwell 1989

First published 1989

Basil Blackwell Ltd
108 Cowley Road, Oxford, OX4 1JF, UK

Basil Blackwell Inc.
Three Cambridge Center
Cambridge, MA 02142, USA

British Library Cataloguing in Publication Data

A CIP catalogue record for this book is available from the British Library

Library of Congress Cataloging in Publication Data

Manufacturing change.
    (Warwick studies in industrial relations)
    Bibliography: p.
    Includes index.
    1. Industrial relations – Effect of technological
innovations on.   2. Technological innovations –
Management.   3. Organizational change – Management.
I. Tailby, Stephanie, 1954–   .   II. Whitston, Colin.
III. Series.
AD6331.M2265 1989   338'.064   89–7275
ISBN 0–631–15983–5

Typeset in 10 on 11-½pt Times
by Vera-Reyes, Inc., Philippines
Printed and bound in Great Britain by
Camelot Press, Southampton

Contents

Contributors

Stephen Evans is Lecturer in Industrial Relations, University of Cardiff.
Roy Lewis is Professor of Law, University of Southampton.
Stephanie Tailby is Research Fellow, Industrial Relations Research Unit, University of Warwick.
Michael Terry is Lecturer in Industrial Relations, University of Warwick.
Peter Turnbull is Lecturer in Industrial Relations, University of Cardiff.
Janet Walsh is Research Officer, Department of Applied Economics, University of Cambridge.
Colin Whitston is Research Fellow, Industrial Relations Research Unit, University of Warwick.

Series Editors' Foreword

The University of Warwick is the major centre in the United Kingdom for the study of industrial relations, its first undergraduates being admitted in 1965. The teaching of industrial relations began a year later in the School of Industrial and Business Studies, and it now has one of the country's largest graduate programmes in the subject. Warwick became a national centre for research in industrial relations when the Social Science Research Council (now the Economic and Social Research Council) located its Industrial Relations Research Unit at the University. Subsequently, in 1984, the Unit was reconstituted as a Designated Research Centre attached to the School of Industrial and Business Studies. It continues to be known as the Industrial Relations Research Unit, however, and now embraces the research activities of all members of the School's industrial relations community.

The series of Warwick Studies in Industrial Relations was launched in 1972 by Hugh Clegg and George Bain as the main vehicle for the publication of the results of the unit's projects, as well as the research carried out by staff teaching industrial relations in the University and the work of graduate students. The first six titles of the series were published by Heinemann Educational Books of London, and subsequent titles have been published by Basil Blackwell of Oxford.

The scale and pace of change in the economy and industrial relations since the late 1970s has posed new problems for analysts in Britain and internationally. Both the causes and direction of change remain controversial, and are reflected in public policy debates and within organizations, as well as in academic research. Attention has increasingly been directed to concepts of restructuring in an attempt to understand the connections between economic, institutional and political developments. Restructuring is not a simple concept, however. The literature is varied, and it is sometimes difficult to disentangle its descriptive, analytical and prescriptive strands. There is a clear need to explore in some depth what is happening in different economies, sectors and firms. Industrial relations studies can play an important part in this, and the multidisciplinary approach of the unit's work is ideally suited to the task.

This book makes an important contribution to debates on restructuring. By looking at different sectors of industry, and dealing with problems from the level of the sector to the shopfloor, it brings out the complexity of the process of change. The case studies provide a wealth of information on how managers and unions have responded to new market conditions and technical innovation. They show the links between market forces and the organization of labour, and question the ability of markets to solve many of the underlying problems of investment and productivity.

The empirical material is placed in a broader theoretical framework which engages with some of the important debates on the future shape of the economy, and of labour relations. The emphasis on social agencies as the determinants of change in particular focuses attention on the state, and explores the contradictory effects of competition between different groups of capital. This book challenges some old, and some emerging orthodoxies, and emphasizes the incomplete and fragile nature of change in manufacturing. It thus makes an important contribution to the debate on restructuring in a number of disciplines including economics and politics, as well as industrial relations.

George Bain
Richard Hyman
Keith Sisson

Editors' Introduction

This book addresses the links between industrial relations and industrial restructuring. It draws together research material from six separate projects conducted at the Industrial Relations Research Unit of Warwick University in the mid-1980s. The projects varied in their principal themes, levels of analysis, and research methods, but each was concerned in some way with the complex process of 'manufacturing change'. By bringing the studies together, we have tried to make a distinctive contribution to the debates on the contours and dynamics of restructuring. Emphasizing the unevenness of restructuring, the book explores the contributions of the state, employers, workers and trade unions to the changing forms and character of economic and social organization in a range of industrial sectors in Britain.

The overall coherence of the book owes much to the discussions which took place at a series of preliminary meetings involving all the contributors. The introductory chapter, which sets out the theoretical context within which the studies are situated, was developed in the light of these discussions. We also wrote the brief introductory notes at the start of each chapter to draw out its key features and to highlight its connection with the theoretical and policy debates outlined in the introduction. We hope that these interventions have not misrepresented the ideas of our contributors.

We are very grateful to the contributors for participating in this project, and for all the effort they put into its development. When the chapters were drafted, the contributors were all resident at Warwick and although the position has since changed, they all retain close links with the Unit. Roy Lewis is now Professor of Law at Southampton University. Steve Evans and Peter Turnbull have both moved to the Cardiff Business School where they are now Lecturers in Industrial Relations. Janet Walsh, who was a PhD student at the Unit, is currently a Research Officer at the University of Cambridge. We would also like to thank Jean Hartley, who is now a Lecturer at Birkbeck College. While based in the Unit, she worked with Stephanie Tailby on a project on 'Organizational Decline and Work-place Industrial Relations'. Material from this project has been incorporated into the study of the materials handling industry in chapter 6.

We are grateful to many colleagues at the Unit for the advice and support we have received in bringing this book together. In particular, we would like to thank Peter Nolan, who has been generous with his help throughout the project, and the editors of the series, who provided helpful suggestions on drafts of the manuscript.

Stephanie Tailby
Colin Whitston

1

Industrial Relations and Restructuring

Stephanie Tailby and Colin Whitston

Industrial restructuring in Britain and, specifically, the contribution of industrial relations to economic and social reorganization since the 1970s are the main themes of this book. Detailed case studies examine the experience in different sectors of British industry, and the role of workers, trade unions, employers and the state. Industrial restructuring is a term that is frequently employed but rarely defined. Indeed, it is 'now so generalized as to virtually have lost its meaning!' (Massey, 1983a). In this account, industrial restructuring has a very specific definition, intimately related to economic crisis.

In capitalist economies there is a compulsion to accumulate; to produce, realize and reinvest profits in new technologies and the production of new commodities. Propelled by competition between firms and the antagonism between employers and workers, the process involves the constant expansion of capacity and its destruction. In this sense it is correct to argue that the closure of plants in declining sectors and the opening of new ones in growth industries is a permanent process, but incorrect to equate this with economic restructuring. The process of accumulation is inherently unstable, or prone to crisis, necessitating a transformation in the conditions for securing the expanded reproduction of capital. Economic crisis is a rupture in the process of accumulation, or a generalized crisis of profitability, requiring for its solution not simply the recovery of existing profitable enterprise, but also the creation of a new basis for growth through the restructuring of economic and social relations.

Economic crisis, in other words, is a turning point in the pace and direction of capitalist development. It is in this sense that writers from a

variety of academic disciplines and analytical traditions have used the term
to characterize the economic and social dislocation experienced in Britain,
along with other advanced industrial economies, since the 1970s. These
accounts have addressed restructuring at the level of the international
economy, for the widening internationalization of trade, finance and
production in the post-war period has meant that the present crisis has
taken on a global character, involving both the advanced and, in different
ways, the developing and centrally planned economies. Nevertheless,
divergent explanations for the collapse of the post-war boom have been
offered and similarly, there have been disagreements as to the elements
required for a generalized recovery. Some of the more influential interpretations are considered in the following section of this chapter. In
these, the origins of the crisis are variously located in the development of a
world market for labour power and production sites; worker wage militancy; and the fracturing of mass markets for mass produced commodities.
The difficulties with each of these interpretations are considered and in
particular, their preoccupation with developments in the sphere of exchange and distribution. A more promising theoretical approach is outlined, based on an analysis of the circuits of capital, and emphasizing the
unity of production, distribution and exchange.

Restructuring is not simply an economic, but also a social process. As
such, far from being fixed in advance, the outcomes are crucially dependent upon the interpretations and practices of social agencies through the
institutions and political organizations mediating the expression of different material and class interests. The particular characteristics and interaction of these agencies have shaped the distinctive patterns of economic
development in different capitalist countries, and the experience of the
present crisis reflects these differences. An important element in the orig-
inality of the British experience is this country's long-term decline relative
to other advanced industrial economies. Against this background, and
focusing upon the role of the state which forms the focus for conflicts
between the major protagonists of capital and labour, the second section of
this chapter examines the politics of restructuring in Britain.

Within the context of these broader economic and political developments, the case studies examine the specific circumstances of industries
located in different positions within the economy – in producer, intermediate and consumption goods sectors. They also cover different levels of
activity within sectors. Two discuss problems affecting entire sectors, while
three look at individual firms against a sectoral background, and one study
concentrates on plant-level issues. This format allows discussion of the
complex interactions involved in restructuring. They explore the ways in
which these sectors have experienced economic crisis and the development
of the relations between firms, between workers, and between workers and
employers shaping the process and outcomes of restructuring at sectoral
and enterprise levels. Each addresses the influence of government policy

on the forms of production and employment reorganization pursued to restore and enhance profitability, and state intervention in restructuring forms a specific focus in the chapters on the foundry, textile, materials handling and construction sectors.

The International Dimension

In this section, two broad interpretations of the nature and origins of the present world-wide crisis are discussed. These perspectives are, of course, structured within wider theoretical interpretations of the nature and dynamics of capitalist development. In the first interpretation, the relocation of productive capital to the low-waged developing countries is viewed alternatively as the cause of, or as a response to, the crisis of profitability in the advanced industrial economies. In the second, changes in consumer demand and the advent of microelectronic technology are held to have reversed this flow of investment, or to have formed the basis for the recovery of profitable production in the high-waged economies of the 'centre'.

Capital Relocation and the New International Division of Labour

Three features of the modern international economy, and the connections in their development, are emphasized in the first account of crisis and restructuring. These are: (1) the increasingly prominent role of multi-national companies, controlling production processes which cross national boundaries; (2) the decline in manufacturing employment as a proportion of aggregate paid employment in the advanced industrial economies, or what has come to be termed 'deindustrialization', and (3) the industrialization taking place in the 'Third World' or 'periphery' countries, formerly integrated into the international economy primarily as importers of manufactured goods from the advanced economies and as exporters of raw materials.

Addressing these developments, writers have given different explanations for the emergence of a new international division of labour, taken to signify the reintegration of the developing countries into the international economy as production sites, sources of labour power and exporters of manufactured commodities. For Fröbel *et al.* (1980), three preconditions have been decisive; (1) the development of a world reservoir of potential labour power; (2) the development of the manufacturing labour process, the decomposition of production processes into elementary tasks and the associated deskilling of the labour force; and (3) developments in the forces of production in the fields of transport and communication rendering production more geographically mobile. These preconditions, together with the profit-maximizing behaviour of firms, are presented as an explanation for the relocation of labour-intensive elements within firms from the 'centre' – the advanced economies – to the cheap labour countries of the 'periphery'. In the latter, the reproduction of labour is subsidized by precapitalist sectors, labour is in abundant supply and largely disorganized, and hence firms can ensure the maximum intensity of use, unconstrained by legislative restrictions on the length of the working day.

In this interpretation, capital relocation to the periphery is seen, in turn, as the cause of the phenomenon of crisis witnessed in the centre – rising unemployment, bankruptcies and mergers, stagnating investment, fiscal problems for the state (Fröbel *et al.*, 1980:47). From a second perspective, however, the relocation of productive capital is identified not as the cause, but rather as a response to the crisis of profitability in the advanced economies, characterized as a 'profit squeeze' and attributed to labour's increasing strength in relation to capital in distributional, or wage struggles (Frank, 1980; see also Glyn and Sutcliffe, 1972). This strengthening of organized labour, which is held to have constrained capital's ability to maintain profitability by increasing the rate of exploitation, is variously attributed to high levels of employment in the long post-war boom, or seen as a secular tendency of capitalist development as a result of the concentration and centralization of labour accompanying the concentration and centralization of capital (Arrighi, 1978). Industrial relocation to the low-waged periphery is therefore identified as the dominant response of employers, facilitating a reduction in the average level of wages on a world scale.

The limitations of these interpretations of economic crisis and the changing international division of labour have been explored by Jenkins (1984) in an excellent critique of both approaches, and only the chief difficulties need be noted here. Focusing almost exclusively upon the capital–labour relationship in the sphere of distribution, the theory of crisis articulated in the second account has been criticized as partial and reductionist. It leads to the projection of a virtually unilinear employer response to failing profitability – relocation to the low-waged periphery – rather than the diversity of strategies that might be anticipated, given the complex formation of industrial and social structures in different national economies. Both approaches identify the search for cheap labour as the prime mover in the process of capital relocation, and suggest that mechanisms to reduce costs through the depression of wages and work intensification are central to capital's need to restore and enhance profitability. While this might strike a chord with, for example, the Conservative government in Britain, it ignores the role of technological innovation in enhancing labour productivity and reducing unit costs. Thus, these approaches neglect the part played by rising wage costs in stimulating the development of firms and industries through investment in labour-displacing technologies.

Certainly, relocation in search of cheap labour has been pronounced in some industrial sectors, notably in the electronics and clothing industries.

But here, as other writers have emphasized, while in the past there have been important constraints on mechanization, new technology has subsequently facilitated the reimportation of these assembly processes to the 'centre' (Elson and Pearson, 1981). Analyses highlighting the role of wage costs in capital relocation and the emergence of a new international division of labour have drawn heavily upon the example of the textiles sector. The study of textiles in this book (chapter 4) suggests, however, the complex web of relations that have influenced one major British multi-national to re-equip its production facilities in the UK, whilst continuing to expand abroad in both the advanced and developing countries, in the pursuit of profitable opportunities. These include the nature of competition in the retail clothing sector, competition from low-wage overseas producers, and divisions within the UK workforce based on gender, shaping trade union responses to technical innovation.

Flexible Specialization
Automation in the 'centre', rather than emigration to the 'periphery', is emphasized in a diferent perspective on contemporary capitalist restructuring. In this approach, the present crisis is seen as a period of transition between a decaying epoch of 'Fordism' and an emergent epoch variously characterized as 'flexible specialization' and 'neo-Fordism' (Sabel, 1982; Piore and Sabel, 1984; Sabel and Zeitlin, 1985). 'Fordism' basically denotes the principles of mass production which supported post-war economic expansion in the advanced capitalist countries – the use of deskilling technologies and 'Taylorist' forms of work organization in the manufacture of standardized consumer goods for distribution to large, undifferentiated markets. Changing patterns of demand are held to have undermined the vitality of this method of manufacturing and, with this, to have rendered the economic, social and political structures of the post-war period redundant. Specifically, mass-markets for standardized commodities are said to have reached saturation in the advanced economies, and demand has become increasingly heterogeneous with the assertion of consumer preference for high quality products and individualized designs.

Developments in the forces of production, in the shape of computer-controlled production equipment, are also identified as conditioning the transition to 'flexible specialization' which involves small-batch production of quality, customized products for discrete or specialized market niches. Effecting time economies in the processing, flow and retrieval of information, microelectronics technology has enabled firms to respond to the changing composition of demand by permitting a reduction in unit costs within non-repetitive manufacturing. Investment in the new technology could be employed to secure economies within 'traditional' mass-production, or for product diversification. Hence, the fracturing of mass-markets must be considered as the decisive development in this explanation of the crisis of 'Fordism'.

'Flexible specialization' is identified as the basis for a capitalist recovery (Piore and Sabel, 1984), and as the new model of competitive behaviour throughout the advanced economies (Best, 1984) and hence one to which firms must adjust. But it is also seen as a model of adjustment that promises to confer on workers significant benefits in the form of enhanced job security and higher wages, more interesting work and improved relations with management. New forms of work organization, therefore, are considered essential to secure the full potential of microelectronics technology and new market opportunities. Thus, the firm's ability to respond flexibly to changing consumer demand is said to be dependent upon the flexible deployment of workers within production. Workers must be competent in a range of tasks, and be prepared to switch between these as the production schedule demands. The recomposition of tasks to attain this flexibility is held to be equivalent to the reskilling of labour. Hence, the transition to 'flexible specialization' is seen as representing a break with the 'Taylorist' traditions of the past (Streeck, 1986).

Task recomposition, it is suggested, raises wage rates and enhances the intrinsic interest of work, and establishes the basis for a new concord between management and labour, which is in turn considered to be essential for enterprise success in the new regime of 'flexible specialization'. For with inter-firm competition becoming increasingly innovation – rather than price-led, managements are said to become more dependent on the creativity and experience of their employees in the development of new products and on their responsiveness in adjusting to rapid changes in product markets. In order to tap this resource, therefore, firms must discard the 'authoritarian' forms of labour control associated with mass-production in favour of participatory industrial relations.

In order to support a product-market strategy based on 'continual inventiveness' (Piore, 1986:211), firms must not only develop new co-operative relations with their employees, but also with supplier companies and with local academic institutions, and act with competitor firms to discourage destructive price-cutting behaviour. Such forms of decentralized, 'community' regulation are considered most appropriate to the regime of 'flexible specialization', the transition to which is in any event held to have lessened the need for macroeconomic regulation on the post-war Keynesian model. This conclusion follows from the argument that the break-up of mass-markets and development of microelectronics technology have rendered the constituent elements of the economy increasingly independent. Effectively lowering the optimal scale of production in former mass-production industries, the new technology is held to have undermined the competitive advantage of monopoly capital and lessened the need for the state to subsidize large-scale, fixed investments. Firms, it is argued, are less dependent upon the state to support mass purchasing power since they can, through the introduction of new technology and the exercise of 'creative entrepreneurial activity', survive and prosper by serving and nurturing specialized market niches. In this interpretation, therefore, scientific progress supplements market forces and the profit-maximizing behaviour of firms to produce outcomes that are beneficial to capital, labour, and society at large, with the demise of monopoly capital viewed as a progressive development, reinstating the sovereignty of consumers.

The similarities between this approach and orthodox economic theory are obvious. Detailed critiques of the 'flexible specialization' thesis have been developed elsewhere (Gough, 1986; Nolan and O'Donnell, 1987; Murray, 1987) and its limitations are pursued in several of the case studies in this book. The identification of certain features of contemporary industrial developments is not at issue, rather it is the explanation given for these developments, and the interpretation of their significance for capital and for labour, taking the latter to embrace workers as consumers, which have been addressed.

Criticism has centred upon the isolation of changing patterns of demand as the cause of the present economic crisis, and the failure to articulate the reasons for these changes. It would clearly be incorrect to suggest that developments within production are ignored. But these are viewed narrowly or in terms of the changing nature and potential application of capital equipment. The introduction of computer-controlled production equipment and 'flexible' working practices are discussed in the studies of the textile industry (chapter 4) and automotive components (chapter 5) in this volume. While the circumstances of these sectors vary in a number of important respects, both the firms examined have reorganized production in response to intense price competition in the context of surplus productive capacity. Hence, rationalization – or disinvestment – is a precondition for the recovery of profitability at the sectoral level and the basis on which 'survivors', or firms remaining in production in these industries, can go on to expand and engage in fresh investment. Technical innovation and work reorganization in the case study firms have been pursued to enhance labour productivity and therefore to transform the social relations of production. But the outcome for workers has been job losses, more oppressive supervision and higher levels of stress. The alleged universally beneficial results of flexible specialization, to say nothing of the idea of the newly autonomous worker, are questioned by such findings.

Crises of Overaccumulation and the Circuits of Capital
The 'new international division of labour' and 'flexible specialization' perspectives have addressed serious sources of instability and change, both within nation-states and within the international economy as a whole, underlying the decay of post-war economic expansion. Focusing upon developments within the sphere of exchange and distribution, however, these perspective have been criticized for presenting a partial and inad-

equate account of economic crisis and contemporary capitalist restructuring. A more promising theoretical starting point has been identified in Marx's analysis of the circuits of capital, emphasizing the unity of production, distribution and exchange (Jenkins, 1984; Fine and Harris, 1979).

A more complete understanding of the collapse of the post-war boom in the advanced capitalist countries from the late 1960s has to take into account the contradictions emerging within and between the spheres of production, distribution and exchange, and between the tendency of the rate of profit to fall and its counteracting tendencies. From this perspective, the crisis of overaccumulation is not simply a consequence of these contradictions but also involves their forcible resolution. In this sense, crisis can be seen as an integral part of capitalist accumulation, rather than as some form of aberration. That is, a crisis is the way in which capital and capital–labour relations are restructured in order to restore (temporarily) the harmonious relationship between the spheres of production, distribution and exchange, and to bring about an increase in the rate of profit.

The most fundamental mechanisms through which this takes place are the scrapping of old techniques and the introduction of new ones, through bankruptcies leading to the depreciation of capitals absorbed by their stronger competitors in a process of centralization, and the elimination of the least productive capacity (Fine and Harris, ch. 5, 1979). It is these mechanisms for restoring productivity that are central to the resolution of the crisis, rather than the simple intensification of work and depression of wages emphasized in the exchange-oriented accounts of the 'new international division of labour'. Hence, periods of crisis are characterized by 'intense competition as individual capitals battle for survival and to establish positions for themselves in readiness for renewed expansion' (Jenkins, 1984:41).

Although originating ultimately in the nature of the wage relationship, the crisis of capitalist reproduction cannot be reduced to conflicts between firms and workers over wages, nor to struggles over the conditions for the introduction of new production techniques. But any resolution of the crisis on a capitalist basis requires the restructuring of the capital–labour relations to reassert the priorities of accumulation. In all advanced industrial economies throughout the post-war period, the state has come to occupy centre-stage in economic regulation, intervening to promote accumulation but nevertheless subject to its laws. Hence, state intervention does not eliminate crisis, but rather politicizes it, forming the focus for conflicts over the basis of recovery. The character of the state's intervention in restructuring therefore reflects the character and state of play of the political forces operating on it, which must be analysed at the level of specific national economies.

The British Economy

Economic and social dislocation in Britain since the 1970s must be under-stood as part of the world-wide crisis, but cannot simply be reduced to it or analysed as a representative sample of the whole. Just as each crisis assumes a specific historical form, so each country experiences it within a specific economic and political framework. An obvious element in the severity of the British experience is this country's long-term relative economic decline.

Deindustrialization and the Politics of Modernization
The total number of jobs in manufacturing in Britain has been contracting for some 25 years, and the share of manufacturing in paid employment has been declining over a far longer period (Massey, 1983b:19). This phenom-enon, however, has not been unique to Britain. It has been a feature of most advanced economies in the post- war period, and has coincided with the years of sustained economic expansion.

The critical issue in Britain has been the ability of a reduced manufac-turing sector to sustain income and investments at a level sufficient to maintain profitability and competitivenes in home and foreign markets. When this is not the case, the economy becomes prey to the dual pressures of imports and capital outflow, which at once increase recessionary press-ures and hinder the reorganization of the economy. Hence, the concept of 'deindustrialization' is related to the productivity and export performance of British industry which have generally failed to match that of major competitors. Moreover, it is not simply that Britain's industrial base has been contracting, but that this country has increasingly assumed a position within the international economy as a specialist in low-pay, low-pro-ductivity industries (Fine and Harris, 1985). Relative decline in the boom meant sacrifices in the rate of growth. In a period of crisis it may represent a critical inhibitor to restructuring.

The debate on the causes of the relative decline has been protracted and sometimes heated. Its roots have often been sought in Britain's imperial past and the structures established in the period of industrial supremacy in the nineteenth century. But this is to underestimate the enormous changes that have taken place in the organization and location of production in Britain and in the world economy and, in particular, the internationaliz-ation of finance and production in which British capital has been involved, with British-owned multinationals now ranking second only to those of the United States in the growth of their international activities (Fine and Harris, 1985). Hence, Britain's current position in the international div-ision of labour must be analysed in relation to these developments in British financial and industrial capital and their influence upon domestic accumulation.

Relative decline and its consequences have produced a perpetual striving for the 'modernization' of the British economy and its manufacturing sector, in which the state has played a crucial role in the post-war period. Within this struggle, all the major social agencies – finance, industry and the labour movement – have sought to secure their interests and aspirations, both within civil society and the state. During the years of the post-war boom, conflicts were played out within a political context often characterized as the new post-war settlement between capital and labour. Its lynchpin was the full employment welfare state, its political expression 'Butskellism', and its political practice a constant process of compromise negotiated through a proliferation of governmental and quasi-governmental bodies which drew the trade unions in particular into a deepening relationship with government. Taken together, this may well be described as the construction of a 'social democratic' state. For, if its outlook and practice where far from being the product of the Labour movement alone, it was the Labour Party and the trade unions which most closely embraced it, and which were in turn most closely associated with it.

While the 'social democratic' state gave expression to the new orthodoxies of managed capitalism, the actual impact of government policy on the structure and performance of the economy was fraught with contradictions. Of these, inflation was the most immediately apparent and most indicative of the general decay of the boom years. The secular rise of state, corporate and personal indebtedness had engendered a long inflationary period, in which recessionary trends were averted by the pledging of future profits to the public deficit. As means of prolonging the boom, however, active fiscal policies increasingly reproduced the crisis conditions they were aimed to avert at a higher level, culminating in the economic novelty of 'stagflation' after the 1974 world recession.

Demand management through fiscal policy weakened the coercive force of the market, although in Britain the nature of and relations between finance, industry and the state failed to provide the mechanisms for imposing direction upon domestic accumulation (Leys, 1985; Hall, 1986). While the level of state subsidy in certain sectors was by no means negligible, in the absence of adequate co-ordination and direction the barriers to reorganization proved increasingly intractable. The history of industrial policy, and the constant recreation of the institutions for intervention, illustrate the fraught nature of this aspect of the state's econ-omic role (Fine and Harris, 1985). Moreover, the role of the unions in economic management, the costs of intervention in terms of welfare spending, and the political accommodation with labour were unacceptable to many British employers.

The corollary of these contradictions was the developing intervention of the state in the conduct of industrial relations and wage determination. With the internal failures of British industry, labour – both within production and through its political organizations – was increasingly identified as the major obstacle to economic reform. At the level of the national economy, an increasingly unmanageable state deficit was attributed to the cost of the 'social wage', and to the ability of unions to defend real incomes against both inflation and government regulation. At the level of the enterprise, union organization was blamed for both wage rigidities and for inflexible work practices. As the crisis developed, forcing public expenditure cuts onto the political agenda, analyses relating increased state intervention, trade union power and industrial malaise attained an increasing status of orthodoxy. In fact, while union influence on economic policy and workplace regulation reached its height in the 1974–9 period, its political and material base was already withering. Apparently too strong to be excluded from economic decision-making, the unions were to weak to develop or pursue their own solutions. The demoralization and division of the labour movement had been effected partly by the 'social contract', under which the trade union leadership pledged its support for cuts in welfare expenditure and in living standards, while the Labour government placed the blame for inflation on the irresponsibility of the unions.

As the political and economic orthodoxies of the boom years began to collapse, neo-liberal groupings within the Conservative Party revived the idea of markets as the most democratic regulator of the economy and consumption. Ironically, the basis of the neo-liberal critique was the assertion of Keynesians that economic management was being increasingly undermined by the action of institutionalized interest groups. The core of the Conservative programme, developed through successive governments since 1979, is that supply will create its own demand if extensive state intervention is abandoned. The aversion to state regulation is not confined to economic policy and industrial relations, but forms part of a total critique of the society that emerged from the post-war years of expansion. Nevertheless, very great emphasis has been placed on the influence of organized labour and the need to reduce its power as a condition for introducing a *laissez-faire* recovery programme.

The Problems of Change
'Rolling back the frontiers of the state', however, remains a heavily loaded concept which tends to disguise the dependence of markets on state power. The exclusion of the unions from political influence, and the general weakening of the labour movement, has required strong state intervention in the form of restrictive legislation and the increased use of the police and judiciary. Government policy, moreover, has further encouraged the export of British capital through the lifting of exchange controls. It has been active in attracting inward foreign investment into Britain, and has played a continuing role in attempting to regulate an increasingly disorganized world market. Hence, state intervention remains central to restructuring , even while the forms of intervention have changed markedly. Nevertheless, while some firms and financial interests have benefited from

these changes, the programme of disengagement remains highly contradictory in its effects on domestic manufacturing and the economy as a whole.

The neo-liberal vision of a revitalized market presupposes that the vigour of a nineteenth-century *laissez-faire* economy can be created within the framework of late twentieth-century monopoly capital. Yet much general experience, and certainly the case studies presented here, indicate that the difficulties of market-led change and absence of effective state intervention in industry remain important barriers to recovery.

The enthusiasm for market-led restructuring tends to underestimate the problems of economic integration which formed the basis for government industrial strategies in the past. Governments of other advanced industrial economies are less willing to rely on market adjustments, especially in fields requiring long-term and expensive research but which are potentially important for product development and productivity-enhancing technical innovations. For example, the battle between the United States and Japan to dominate the 'fifth generation' of computer technology has required their governments to muster the forces of entire economic sectors, to co-ordinate research, and protect fledgling products and future markets. This contrasts sharply with the experience in Britain where the Government's failure to intervene at the microeconomic level has been criticized in the Aldington Report (1985) on the decline of manufacturing, and where the Government has been accused of an 'anti-industry' bias by the heads of two of Britain's largest firms (*Financial Times*, 26 April 1985).

The Conservative programme of 'disengagement' corresponded well to the initial, destructive phase of crisis, effecting the break-up of institutions and policies that had characterized an increasingly sclerotic economic structure. But its efficacy as a programme for restructuring has been consistently challenged at both a theoretical and an empirical level (e.g. Leys, 1985; Lipietz, 1984). Thus, Britain's escalating balance of payments deficit indicates the continuing difficulties of domestic manufacturing and the failure of Government policy to redress 'deindustrialization'. This failure represents enormous difficulties for the owners of industry, as well as for the working class, as even multinational companies cannot sever their links entirely from the domestic economy.

To develop a full analysis of the problem of restructuring, attention must be paid to the objective requirements for revitalized accumulation, and to the extent to which these are addressed by the advocates of a market-led recovery. Restructuring takes place on a number of levels and hence is subject to a markedly uneven pace of development. Within a long period of depression, therefore, there are smaller cycles of recession and recovery reflecting this unevenness. Broadly speaking, however, restructuring on the basis of capital must entail a number of elements. Firstly, there must be a drive to reduce labour costs in which the reduction of workers' real incomes can play only a limited role. The critical issue is the utilization of labour power, with the introduction of new production techniques and

forms of work organization securing a reduction in unit costs. Nevertheless, there must be a general redistribution of revenue from labour to
capital as a whole, to which the expulsion of labour from manufacturing
contributes. Also involved are cuts in public welfare expenditure and in
wages in public sector employment. Low-paid employment, particularly in
the growing service sector, takes advantage of unemployment and new
sources of labour supply, produces profits and lessens the impact of
reduced demand. While the Conservative government's policies have
focused on the depression of real wages in all sectors of employment,
however, in the long run the dependence on a low wage workforce
represents a real barrier to recovery, unless new markets for commodities
can be rapidly developed. Moreover, and as is exemplified in the construction case study (chapter 3), the availability of cheap labour may inhibit
the introduction of productivity-enhancing forms of work reorganization.

Secondly, restructuring must reduce the costs of investment, principally
by raising productivity in the production of capital goods. Computer
technology, with associated robotics and design advances are crucial here,
but what is most striking is the relative slowness with which these new
forces of production are emerging. Involved here are both market restraints on the production of consumer goods and the difficulties where the
scale of capital required for the effective utilization of the new producer
goods remains high compared to the profits available under older competitive conditions. Several of the case studies illustrate how the nature of
existing industrial structures can retard the development of new technologies where market forces alone have been insufficient to produce
rationalization.

Thirdly, restructuring must reorder both production techniques and
products in the consumer goods sector in order to create new market
opportunities. Despite the growth in some new consumer goods few
developments in recent years seem to represent the coherent outlines of a
new basis for expansion comparable with the 'new industries' of the inter-
war period. In both the producer and consumer goods sectors, in fact, it
seems probable that substantial advances will make 'technological unemployment' more of a problem of recovery than of the recession of the early
1980s.

Fourthly, the internationalization of the economy and its extremely
uneven development require a new capitalist hegemony along the lines
played by the United States in the crucial period of post-war reconstruction. For the restriction of markets and accelerated indebtedness of the
developing countries are the essential counterpoint to increased instability
in the relations between the advanced economies, financial instability and
trade rivalries.

Finally, recovery relies on the consent – or at least acquiescence – of the
working class in the process of change and the nature of the society that
emerges from it. Unemployment and the weakening of the trade unions

have to date favoured the reorganization of labour markets and changes in the use of labour in production. There is no guarantee, however, that this apparent passivity can be maintained for years into the future.

The ascendancy of neo-liberalism in Britain has played a critical role in breaking up the political consensus of the post-war years. Nevertheless, state intervention remains vital for industrial restructuring as the preceding comments have made clear. The exclusion and weakening of the trade unions may have formed the basis for a political change of course and the return of industrial policies, with intervention untrammelled by the necessity to co-opt union leaders into the process. But much is dependent upon the demands placed on the state by the various sections of British industry and finance. The history of state intervention through the 1960s and 1970s suggests that there are more fundamental barriers to its efficacy than the enhanced political influence of the unions cited by neo-liberal critics (see Leys, 1985; Hall, 1986).

The Experience of Change
The return of the Conservative government in 1979 signalled a significant break with the political culture of the post-war boom and the case studies span this period of transition. They explore, at a more disaggregated level, the forces of change as experienced in different industries and examine the nature and interaction of the social agencies shaping responses at sectoral and enterprise levels. No claim is made for 'typicality' in these case studies. The notion of the 'ideal typical' runs contrary to the argument developed through this chapter and belies the nature of restructuring as a social process. Hence, the studies illustrate the partial and differentiated nature of industrial restructuring in Britain and give a clearer understanding of the reasons for this experience.

State intervention and competition. The process of competition has to be understood against the background of state intervention and broad economic policy as it has developed over the period. The impact of these interventions, however, varies markedly across and within different industrial sectors. Thus, the supposedly virtuous influence of market forces actively fostered by the present government has to be set against the particular economic and social structures governing the development of different industries.

For more fragmented sectors of the economy, for example, and in particular for intermediate producers, competitive rationalization faces important institutional and structural constraints. Firstly, the logic of rationalization is unacceptable to many smaller firms which can remain in production, employing backward production methods, in spite of intense price competition. The more efficient producers may quit competition in that sector and reinvest in other areas promising more immediate profits. Secondly, smaller firms are dependent to a large degree on the investment strategies of larger companies and multinationals, and may also be squeezed by the pricing practices and market power of their suppliers. In the short term, this may serve the interests of buyers and suppliers, but the consequences make it more difficult for small firms to respond to changing demands in quality and product development.

Such was the experience in the foundry industry (chapter 2), which prompted state intervention in the first instance. Fragmentation in capital ownership delayed investment in advanced plant and led to destructive price competition. Government action in the 1970s aimed to modernize and concentrate the industry, but was frustrated by both conflicts of capitalist interest within the sector and the failure to co-ordinate development with end user sectors. Under the Conservative government, an initial return to dependence on market forces proved no more effective, leading the government to back a privately organized crisis cartel.

Restructuring requires the harmonious development of capital and consumption goods industries, and the failure of the market to achieve this underlay much of the government industrial strategy in the 1970s. But the pattern of intervention was often narrowly focused on individual industries and indicative planning did little to resolve problems of inter-sectoral co-ordination or the unwillingness of firms to quit competition. These twin constraints to reorganization are illustrated in the case of the materials handling company (chapter 6). In the 1970s, government used its ownership of the company to encourage mergers, but the domestic market was too narrow to support reorganization on a scale large enough to compete with foreign firms in an increasingly overcapitalized international market. A partial restructuring of ownership within the European industry has developed but with persistent overcapacity, the future for the British firms within the growth area of automated handling systems is currently uncertain.

In both of these examples the lack of a strategic approach to government intervention has combined with the destructive effects of competition to inhibit restructuring. Such outcomes, however, are not universal. In the textile industry (chapter 4), the activities of the leading companies and government intervention throughout the post-war period have combined to produce a highly concentrated sector dominated by three multinational producers. Despite the ambiguities in government policies throughout this period, intervention has been vital for the company studied, which as a multidivisional, multinational concern has been able to alter the nature and location of its investments in response to changed competitive conditions in the 1980s.

Competition and government economic policy have combined in yet another way in the construction sector (chapter 3). The industry has been directly affected by macroeconomic policies which have used it to regulate aggregate demand, and by changing welfare priorities. Resulting uncertainties have encouraged a differential form of industrial structure in which a small number of large companies dominate a galaxy of smaller firms. Combined with the deregulation and privatization policies of the Conservative government, this has produced a highly fragmented industry, with a casual and undertrained labour force employed in low cost, but low productivity, production.

The competitive advantage of large firms. While market forces have failed to promote reorganization in the more fragmented industries studied, the examples of what might be termed relatively successful restructuring concern the textile and automotive components companies, both multi-divisional, multinational corporations with a forceful position in their respective markets. In the case of textiles, as suggested, government intervention in the post-war period has played a significant role in transforming what had formerly been a fragmented sector, apparently in long-term decline. This has in turn enabled the case study firm to respond flexibly to intensified international competition in the 1980s and to co-ordinate the reorganization of its activities internally.

Following the closure of capacity in the UK and the concentration of production on to fewer sites, the firm has invested in new technology to reduce unit costs whilst extending its product range to maintain and develop market opportunities. This example indicates that innovation and market responsiveness are by no means the prerogative of small firms, as suggested in the 'flexible specialization' thesis. Rather, the features of this particular company have enabled it to pursue a variety of responses to maintain and enhance profitability, including its continuing expansion internationally.

The study of the automotive components company (chapter 5) suggests that the changing industrial structures required to establish the new 'flexible' manufacturing systems lead to the increasing centralization of capital ownership and control. With intense competition in this sector, and the run-down of the British motor vehicle industry, the firm has been forced into retrenchment and the search for new markets. Competition between motor vehicle manufacturers, moreoever, has compelled the firm as a parts supplier to pursue both product and process innovation. It has introduced 'just-in-time' (JIT) production, a system which combines new forms of labour control within production with new relations of vertical dependence between buyer and supplier firms. The move towards single-supplier JIT purchasing agreements in the motor industry favours large component manufacturing firms which are also required to assume much of the burden for new product development. To secure the effective operation of JIT, these firms have in turn to exert tighter controls over the internal operation of their supplier organizations to absorb demand fluctuations. Hence, the increasing interdependence of industry is established in contrast to the portrayal of heightened managerial autonomy in the 'flexible specialization' thesis.

Both these examples raise questions, moreover, about who gains and who loses in the process. In the textiles firm, union co-operation played an important part in influencing the location of investment in the UK. The union's response has, however, to be set in the context of high levels of unemployment in the industry and divisions within the workforce based on gender relations. Similarly, the extended 'vertical dependence' between firms associated with JIT production requires the use of subcontractors' employees as relatively low-paid 'shock absorbers' in ways which can create a disadvantaged 'submarket' for workers outside the larger firms.

There are also negative aspects to this restructuring within the succesful companies. Process innovation in both the textiles and automotive components companies has restructured capital–labour relations within production, but not in the manner envisaged in the more optimistic accounts of 'flexible' manufacturing. 'JIT' involves task recomposition which intensifies work by undermining established union job controls and rendering management the sole authority over labour deployment. New labour management practices have been introduced in order to maintain these heightened levels of work effort by substituting group pressures for direct supervision and securing the internalization of control by employees themselves. While the effect is to undermine established forms and practices of workplace union organization, as an 'ultra taut' production system, JIT also renders managements increasingly vulnerable to the withdrawal of co-operation by workers either internally, or at supplier or customer companies.

Trade union organization. The experience of change since 1979 has created a major challenge for the trade unions. Organized labour has been identified as major element in industrial decline by the Conservative government, which sees both its political influence, and its shopfloor organization as barriers to its *laissez-faire* recovery programme. This analysis has already been challenged, and the studies that follow emphasize the relative weakness of unions in this period, and question the idea that competitive success on capital's terms constitutes a universal interest. On the contrary, the demobilization of labour has not only placed many of the burdens of restructuring on the shoulders of working people, it has also contributed to partial and contradictory forms of restructuring. This raises some important questions about the role of trade union organization as an agency of change.

Throughout much of the period covered in this book, unions were closely identified with interventionist strategies for modernization, but the results proved disappointing. The attempt to build a coherent and co-operative approach to change was insufficient to overcome the distortions of intercapitalist competition and the relatively ineffective role of the state. The contradictions involved in the assumption of common interests between workers and employers were graphically illustrated in the report of

Four Trades Councils (1982) on the industrial strategy of the 1974–9 Labour government, and are echoed in many of the present studies, which also explore the limitations imposed on union influence by the gap between political aspiration and industrial organization.

In the case of foundries, the fragmented structure of capital ownership was replicated in fragmented union organization which constrained labour's ability to develop its own solutions for the industry's difficulties. The unions were strong supporters of the NEDO initiatives in the 1970s, and in some ways showed a surer grasp than did the employers of the need for structural reorganization. Yet their commitment to negotiated change within the existing sectoral framework left them unable to mobilize their shopfloor organization or to develop the necessary connections with customer and supplier industries. A similar paralysis affected union intervention in materials handling, where co-operation with employers not only prevented the development of an independent approach, but tended to collapse into the defence of individual firms during a period of rising unemployment.

This dislocation between strategic aspirations and immediate responses to competitive pressure has served to undermine shopfloor responses to change. Although much survey data attest to the survival of workplace union organization and industrial relations structures within industry, diverse conclusions can and have been drawn from this evidence, including the continuing importance of shopfloor organization for the management of change. Certainly, the scale of economic and social dislocation throughout the economy raises a number of questions relating to the nature and limitations of British workplace union organization.

The unions have been unable to exert much influence on either the pace or direction of change since the late 1970s, often accepting a managerial logic of change as a response to seemingly irresistible external competition. The study of reorganization in the pharmaceuticals and food processing industries (chapter 7) emphasizes the part played by product markets, rather than the level of unemployment, in shaping shopfloor union responses, and how they interact with the 'factory consciousness' that has characterized much post-war workplace organization. Workplace organization has had little to do with what gets produced, or how it is produced, concentrating essentially on the effort bargain. In a period of intensified competition, this may be translated into a positive identification with company aims, or at least an unwillingness to challenge them. Where the active co-operation of the workforce in production has been enlisted, it has often been a crisis measure, as in the case of the materials handling firm (chapter 6), which in this instance proved insufficient to arrest the company's deteriorating economic performance.

If the structures of workplace industrial relations remain intact, however, this does not mean that substantive power relations are unchanged. The 're-establishment' of managerial prerogative and the downgrading of collective bargaining have in many cases undermined the status of shop stewards, and a new emphasis on the 'individualized' employment relationship has been carried over into work organization and the choice of technologies. The JIT production system in the automotive components firm also effected important changes in the nature of management–labour relations, and relations between workers, which have undermined workers' collective ability to influence change. Similar conclusions are reached in the studies of textiles, materials handling, pharmaceuticals and food processing.

This weakening of labour, is, however, a double-edged weapon for management, encouraging a recovery of profitability on the basis of relatively restricted productivity improvements, as the case of the construction industry demonstrates. Deregulation and the casualization of the labour market in the 1980s has further weakened the unions and divided the workforce with the diverse forms of employment contract associated with subcontracting. These divisions and the nature of union organization have militated against labour's ability to act as a force for the reorganization of the industry through the introduction of productivity-enhancing changes in production techniques.

In contrast, Streeck (1986) has argued that the ability of Swedish and German car companies to introduce flexible production systems rapidly and effectively has been heavily influenced by the presence of unions in the workplace and in the polity arising from legal regulation. Union pressures have encouraged firms to adopt product and process innovation which raise productivity and develop new markets, and to find means to use existing workforce skills effectively. The result, in terms of labour's co-operation in production and the development of workplace relations, is contrasted with Britain where managements have been able to undermine workplace organization and to employ the coercion of the external labour market to secure change. In the process, British firms may have relinquished a valuable productivity resource in terms of the social relations within production.

Care must be taken with this argument. The implied benefits of flexible production systems for labour have to be viewed in the context of a system which remains exploitative, contradictory, and potentially antagonistic. Also, capital structures and industrial relations institutions have to be viewed in their national context and their place in the international market, and the optimism Streeck expresses may be highly dependent on growth levels. Beneficial changes in labour organization and productivity have, historically, proved vulnerable to renewed competition, and thus subject to renewed pressures for labour intensification. In a relatively weak manufacturing economy such as Britain's, the negative effects on the workforce of restructuring mean that the shopfloor 'truce' is liable to prove highly unstable. Nevertheless, the implication is that the assertion of management's unfettered 'right to manage' – whilst effective in the short term as a means to secure change – may in the longer term impose serious constraints on British managements and British industry.

The case studies address themselves to these complexities and uncertainties. While each examines in detail some aspect of the problem of restructuring, taken together they also show how partial are many of the prescriptions on offer and the solutions adopted. The fragile nature of the recovery after 1982 emerges strongly in all the studies, drawing attention to unresolved contradictions at the social and political as well as economic levels. Restructuring is thus seen as neither an automatic, nor a socially neutral process. While tracing the breakdown of both the economic and political characteristics of the boom years, these studies question the new enthusiasm of those who see a brave new world emerging from the operation of market forces, as well as examining the shortcomings of previous attempts at economic regulation. Thus, in contrast to both neo-liberalism and the proponents of flexible specialization, the argument for concious social control over the means and aims of production reappears in this analysis, and is related to the distinct interests of social classes in the outcome of restructuring. Much remains to be done in this respect, but this work hopefully will form a contribution to further research and debate.

Bibliography

Aldington, Lord. 1985. *Report from the House of Lords Select Committee on Overseas Trade on the Causes and Implications of the Deficit in the UK Balance of Trade in Manufactures*. HL 238/1. London: HMSO.

Arrighi, G. 1978. 'Towards a Theory of Capitalist Crisis', *New Left Review*. No. 111, September–October, 3–24.

Best, M. 1984. 'Strategic Planning and Industrial Policy', *Local Economy*. Vol. 1, 65–77.

Elson, D. and R. Pearson. 1981. '"Nimble Fingers Make Cheap Workers": An Analysis of Women's Employment in Third World Export Manufacturing', *Feminist Review*. No. 7, 87–107.

Financial Times. 1985, 26 April.

Fine, B. and L. Harris. 1979. *Rereading Capital*. London: Macmillan.

——. 1985. *The Peculiarities of the British Economy*. London: Lawrence & Wishart.

Four Trades Councils. 1982. *State Intervention in Industry*. Coventry, Liverpool, Newcastle and North Tyneside Trades Councils.

Frank, A.G. 1980. *Crisis in the World Economy*. London: Heinemann.

Frobel, F., J. Heinrichs and O. Kreye. 1980. *The New International Division of Labour*. Cambridge: Cambridge University Press.

Glyn, A. and R. Sutcliffe. 1972. *Bristish Capitalism, Workers and the Profit Squeeze*. Harmondsworth: Penguin.

Gough, J. 1986. 'Industrial Policy and Socialist Strategy: Restructuring and the Unity of the Working Class', *Capital and Class*. No. 29, Summer, 58–81.

Hall, P. 1986. 'The State and Economic Decline' *The Decline of the British Economy*. Ed B. Elbaum and W. Lazonick. Oxford: Clarendon.

Jenkins, R. 1984. 'Divisions Over the New International Division of Labour', *Capital and Class*, No. 22, Spring, 28–57.

Leys, C. 1985. 'Thatcherism and British Manufacturing: A Question of Hegemony'. *New Left Review*, No. 151, May–June, 5–25.

Lipietz, A. 1984. 'Imperialism or the Beast of the Apocalypse', *Capital and Class*, No. 22, Spring, 81–111.

——. 1983a. 'Industrial Restructuring as Class Restructuring: Production Decentralization and Local Uniqueness', *Regional Studies*. Vol. 17, No. 2, 73–89.

Massey, D. 1983b. 'The Shape of Things to Come', *Marxism Today*. April, 18–27.

Murray, F. 1987. 'Flexible Specialisation in the "Third Italy" ', *Capital and Class*. 33, Winter, 84–95.

Nolan, P. and K. O'Donnel. 1987. 'Taming the Market Economy? An Assessment of the GLC's Experiment in Restructuring for Labour', *Cambridge Journal of Economics*. Vol. 11, September, 251–64.

Piore, M. 1986. 'The Decline of Mass Production and Challenge to Union Survival', *Industrial Relations Journal*. Vol. 17, No. 3, 207–13.

Piore, M. and C.F. Sabel. 1984. *The Second Industrial Divide: Possibilities of Prosperity*. New York: Basic Books.

Sabel, C.F. 1982. *Work and Politics: The Division of Labour in Industry*. Cambridge: Cambridge University Press.

Sabel, C. and J. Zeitlin. 1985. 'Historical Alternatives to Mass Production: Politics, Markets and Technology in Nineteenth Century Industrialization', *Past and Present*. No. 108, August, 133–76.

Streeck, W. 1986. 'Industrial Relations and Industrial Change in the Motor Industry.' Public Lecture, University of Warwick.

Foundries

The transition from government intervention to 'disengagement' is the subject of this study. Facing long-term decline, the foundry sector nevertheless has an important role as a producer of intermediate goods. Intervention in the 1970s, via the Foundries Assistance Scheme and Foundries Economic Development Committee, (FEDC), aimed to cure problems of underinvestment and fragmented ownership. This programme failed in the context of materials substitution and developing overcapacity. The major obstacles to restructuring were the divergencies of capitalist interest in the industry and the near monopoly power exercised over it by supplier and customer companies.

After 1979, the Conservative government refused further assistance, attributing the industry's difficulties to the distortion of markets as the result of previous interventions. Market forces, however, proved no more effective than hitherto and the government backed a crisis cartel organized privately by a merchant bank. The aim was to accelerate market-led restructuring by promoting the interests of large firms in the sector. The cartel failed.

A central focus on the study is the role and aspirations of trade unions and their involvement in intervention. The unions had high expectations that this would improve conditions and secure stable employment, and of all the participants were the most committed to a strategy of modernization. Their ability to secure change on these terms was undermined by fragmented union organization and the collaborative structures of tripartism. The study critically examines union approaches to industrial planning and concludes that a narrow, sectoral approach is insufficient, given the position within engineering occupied by the foundry industry.

2

Rationalizing Foundries

Colin Whitston

The relationship between the state, unions and employers during a 10-year period of decline and rationalization in the ferrous foundry industry from 1975 to 1985, is the focus of this study. The opening of the period coincided with the adoption by the Callaghan government of the 'new industrial strategy' which, for the unions at least, represented an attempt to couple indicative planning and collective bargaining at a sectoral level. Yet during these 10 years neither 'traditional' forms of industrial organization nor political participation seemed to offer the unions any hold on events. An analysis of the failings in the industry therefore demands some questioning of the principles and practices that have for some time underpinned union policies on economic management and industrial strategy.

State intervention was aimed at encouraging the reorganization of a fragmented industry racked by overcapacity, technical inefficiencies, and financial difficulties, and increasingly threatened by import penetration. In the process a variety of means was used, ranging from direct subsidy through the tripartite work of the Foundries Economic Development Committee (FEDC), to state sponsorship of a crisis cartel organized by merchant bankers, Lazard Frères. These differing approaches in turn represented differing political circumstances and aims, as well as differing understandings of the nature of the developing crisis within the industry.

The trade unions were initially enthusiastic supporters of change, seeing in a modernization programme both safeguards for jobs and the chance to improve conditions in an antiquated industry. As decline became more serious, acceptance of change became less a matter of optimism than the need to protect as many jobs as possible. Yet, whether in hope or desperation, the assumption persisted that there was a way in which change could be managed in the interests of the employees, and that organized labour within the industry, and through its relations with the state, could secure that outcome.

There were two levels of problems which undermined these assumptions. First, the nature of planning advocated by the unions proved highly bureaucratic, with little or no connection between the activities of the senior officials and the conduct of union business on the ground. This gap, and the collaboration with the employers required by tripartism, led to a paralysis in the face of capital's own estimation of the nature of restructuring required.

The second problem lay in the structure of capital ownership and competition. The industry is highly fragmented, and in orthodox economic terms very competitive. Fierce competition has, however, produced only destructive results, with ruinous price cutting, and the inhibition of change. Many of these problems are the result of the relationship of small and medium-sized capitals in the industry to their powerful suppliers and customers. It will be argued that it is the nature of conflict between capitals that has most shaped the fate of the sector, and that the realization of the unions' aspirations cannot be resolved by action limited to the sector itself. This has important implications for restructuring in general.

The industry itself cannot be seen as a site for a general restructuring of profitable production. As a producer of intermediate and final commodities, it is highly dependent on developments elsewhere, and has neither the innovative potential nor the market power to play such a role. It will be a consumer, rather than a producer, of new technologies and products. To say this does not mean that the industry is not important for the general development of the economy. On the contrary, its skills and products, on however reduced a scale, will have a part to play in the regeneration of manufacturing. A successful restructuring of British capital depends not only on the development of new manufacturing techniques, products and markets by the leading producers, but also on developing the less dramatic but equally essential sectors that support and complement them. It would be as if the electric power industry could develop without the electric lamp. The fate of the industry in the late seventies and early eighties is, moreover, in many respects symptomatic of the barriers experienced by British capitalism in restructuring, and of the complex role of employers, government and unions in this process. Although in some ways an extreme case, the industry usefully exemplifies problems of planning and restructuring because it displays the essential tensions particularly starkly.

The Structure of the Industry

An overview of the ferrous sector is difficult to provide without reference to other parts of the industry, partly because of the nature of available statistics, and partly because of overlapping markets and institutions. Here, and in the discussion of tripartism, the whole industry is referred to, but the emphasis remains on ferrous foundries which constitute its largest part, and which share (often in an exaggerated form) the general problems of the industry.

In principle, the scope of the foundry industry is simple to define, but in practice the conventional description under SIC Order VI can give a misleading aura of integrity to what is in fact a fragmented industry. For collective bargaining purposes, for example, foundries are divided by weight into Engineering (heavy castings) and the Light Castings Industry. Another division may be made on the lines of process type, which differentiates, for example, between die casting, sand casting, and investment (lost wax) castings. Again, distinctions may be made between foundry types: jobbing foundries (one off, short runs), tied foundries, and volume production repetition castings. Such a multiplicity of possible subdivisions of a small industry indicates not only its fragmented nature, but also its division into largely non-competing sectors.

Employment figures are difficult to calculate due to changing definitions and counting practices. Taken together, however, figures from the Foundry Industry Training Board (FITB), the Department of Employment and the Amalgamated Union of Engineering Workers (Foundry Section) (AUEW (F)) show very similar trends. FITB figures are favoured here because they not only cover establishments exclusively concerned with founding, but they also provide a breakdown by region, establishment size and occupation.

The picture that emerges is of an industry composed of a small number of establishments whose numbers remained steady through the late sixties and early seventies, but began to decline in 1976–7. FITB figures show a decline in establishments from round 1,500+ in 1969 to 1,331. Establishments show a marked variation in size and distribution. In particular the period has seen a marked shift away from large foundries employing 1,000+ people, in favour of smaller plants, falling from 1.4 per cent of the industry to 0.6 per cent.

While employment was growing at the start of this period, its decline began earlier than that of establishment size. Total employment in 1968 stood at 147,000, but fell by 32 per cent, to 99,676 by 1980. Most important was the marked fall in the percentage employed in the largest category of foundry size from about 25 per cent in the early seventies to about 11 per cent in 1981. Distribution by occupation did not change dramatically, although management and supervisors take an increasing proportion at about 10 per cent. Only scientists and technologists increased absolutely, but they still accounted for only 0.5 per cent in 1980.

Because foundries in general provide primary / intermediate products to industrial customers, they are vulnerable to substantial demand fluctuations over any business cycle. A study by the West Midlands Economic

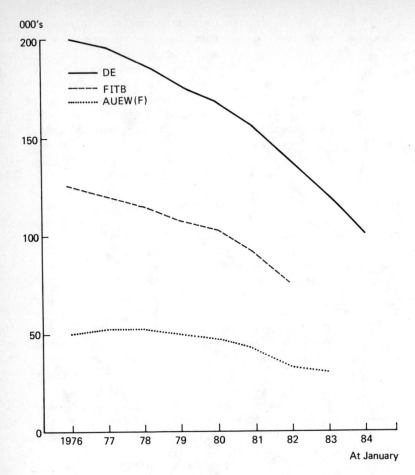

FIGURE 2.1
Foundry Workers in Employment, 1976 – 1984 (000s)

DE – Department of Employment
FITB – Foundry Industry Training Board
AUEW (F) – Amalgamated Union of Engineering Workers (Foundry Section)

Development Unit (WMEDU) has indicated that changes in demand for castings are also often greater than changes in economic activity as a whole. The fluctuations shown for iron castings in figure 2.2 are not untypical. While many factors influence the industry (see below), a central problem has been the decline of its major industrial customers. The most dramatic decline originated in the steel industry where demand for ingot moulds fell by over 64 per cent between 1975 and 1980. Over the same period demand for ferrous castings in the automotive industry fell by over

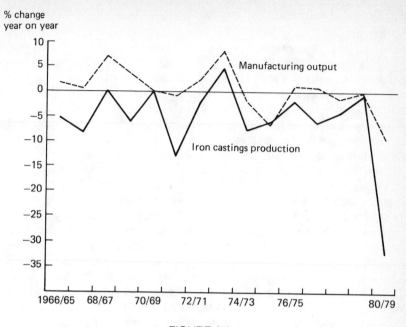

FIGURE 2.2
Output Fluctuations in Manufacturing and Iron Foundries

Source: Economic Trends and The Council of Iron Foundry Associations

38 per cent, and for aluminium castings by over 21 per cent. Building, rainwater goods and general engineering, all major consumers of ferrous goods, showed declines of between 22 per cent and 46 per cent (WMEDU, 1982: 8,9).

While closures and job losses accelerated in 1980–1, the decline in the sector is clearly over a much longer period. The number of iron foundries declined by 22 per cent between 1975 and 1980. Iron output fell by over 50 per cent in the same period, and in steel the fall was 45 per cent (WMEDU, 1982: 13, and *Engineer*, 26 May, 1983: 22–3).

While this general picture of decline undoubtedly finds its roots in the position of the major consumers of foundry goods, the nature of that decline is far from simple. It is already clear that the industry has not reacted to recession in ways that would be normally expected, that is by a concentration of the industry in the hands of more centralized and more profitable plants, and by the elimination of marginal production. On the contrary the problem has been viewed as how 'not so much to improve the healthy as to eliminate the weak' *Management Today*, (December, 1981:35).

The reasons for this failure to adjust are important. In their report on the Lazard scheme Baden Fuller and Hill argue that the problem consisted of an unanticipated collapse of demand aggravated by tardy and inefficient market adjustments (1984: 3,5). This is a reasonable description, but hardly a basis for a satisfactory analysis. The development and the trajectory of the sector's overcapacity crisis needs to be set in a wider context which can show how such market failures are possible.

The competitive sector of manufacturing industry does not operate as an autonomous branch of the economy as a whole. On the contrary, very many such industries are, in effect, clustered around one or more highly concentrated customer sectors, or are crucially dependent for their operation on monopolistic suppliers. The impact of both these factors is very high in the case of foundries. If enterprise decisions are to reflect (and anticipate) shifts in demand in this context, a core of information is required on which to plan investment and output. It is in this area that fragmented industries like foundries are naturally deficient. In effect, pressure for change has therefore occurred not only through the market, but has come also from customers and from political decisions. As a result, the structure of the industry's problem has largely been determined externally. 'Competition' between foundries therefore takes place on terms dictated by far more highly organized sections of capital, and their commercial and political influence. At the same time, the information required to deal with these pressures has been collated largely by public rather than private action. A brief discussion of market perspectives will therefore assist consideration of the part played by these external agencies.

Despite work done by the Steel Castings Research and Trade Association (SCRATA), and some of the larger concerns such as Birmid Qualcast, serious attention to structure of demand for the industry as a whole was the responsibility of the Foundries Development Committee (FEDC) and of the WMEDU. In particular, the FEDC produced three important reviews of market perspective in 1983 which inform much of the following analysis. Survey findings indicate a continuing, if less dramatic decline over the period 1985–90, with total cast tonnage expected to settle at between 30 per cent and 49 per cent of 1975 output. Of individual sectors, only aluminium is expected to regain up to 90 per cent, and investment castings to exceed 1975 output, on the basis of technical changes in customer industries. The dominant ferrous sector is not expected to exceed 37 per cent of 1975 tonnage.

More importantly, the FEDC tried to assess how far this decline was a factor of recession in customer industries, and how much was due to technical and structural reasons. They concluded that: 'Much of the forecast decline is owing to the continuing displacement of castings for technical reasons and will occur even if UK economic activity revives and feeds into UK customer industries' (FEDC, 1983a:4). They cite, in particular, the shift from electromechanical to electronic machinery. The FEDC findings are summarized in table 2.1. In addition, substitution for castings (mainly by fabrication) is complemented by substitution *within* the

TABLE 2.1
Factors Affecting Demand

Main Type of Forecasting Factor	Proportion (%) of 1975 UK Foundry Tonnage
(a) Economic decline of the customer industry	14
(b) Both (a) and (c)	38
(c) Technical substitution for castings	48
Total	100

Source: EDC FS (83)21 (unpublished)
Note: From 42 castings categories but excluding automotive castings

sector – for example, aluminium alloys for iron, spheroidal graphite (sg) iron for grey iron – which have important implications for the structure of the industry. There is also a longer-term threat posed by developments in non-metallic, cintered ceramics being developed in Japan and elsewhere.

The problem of demand therefore has a dual element of customer recession and structural changes that raises many more problems for the industry. These include costs and financial performance, capacity utilization, casting quality, technology and imports.

Increased costs of production were identified by the WMEDU as a major perceived problem in the industry, due to the effects of overcapacity and declining labour productivity, and the rising costs of bought-in materials, components and energy. Wage levels are not generally considered a problem. The main complaints are laid at the doors of nationalized industry, and central and local government.

> Our rates keep increasing, the Electricity and Gas Boards seem to be oblivious . . . and continue to increase prices. Nationalised and monopoly suppliers upon whom we are completely dependent realise their position of strength and imposed increases from which we cannot escape (WMEDU, 1983a: 5).

Table 2.2 shows both areas of rapidly rising costs, and an indication of finished price movements.

The coincidence of demand and cost pressures impacted on financial performance to a critical extent. 1982 figures from Inter Company Comparisons (ICC) sources rank ferrous foundries 56th, and non-ferrous foundries 76th out of 77 sectors, taking profits as a percentage of assets, capital employed, and sales. Return on capital in iron declined from 17.1 per cent in 1977–8, to 6.2 per cent in 1979–80, and in non-ferrous foundries

TABLE 2.2
Costs and Prices

	Percentage Increases in Costs and Prices from January, 1979 to June, 1982
Energy and Materials	
Carbon dioxide	54
Coal dust	87
Oil	132
Sand	65
Liquid oxygen	46
Electricity	98
Gas	71
Rates	103
Labour	20
Metal	20
Selling prices (random items)	
1	8.6
2	15
3	5.9
4	6.4
5	11
6	− 4.25
7	− 10
8	No change

Source: The West Midlands Foundry Industry, WMEDU, p. 15

from an already low 4.1 per cent to a negative − 1.2 per cent (WMEDU, 1982: 16,17).

Given these factors, capacity utilization, and capacity quality also become more crucial. The contraction of enterprises and employment has by no means solved the problem of overcapacity, and the FEDC concludes that, overall,

'There seems to be sufficient capacity to cope with any likely demand revival. The total of 'previous best years' is only 10 per cent higher than that of current capacity on reduced manpower levels and 14 per cent less than current capacity with labour taken back on (FEDC, 1983b: 4).

As a result, the FEDC presented three important conclusions. Firstly, a high main batch size (over 10,000) is associated with the highest utilization. Secondly, utilization increases with the size of the foundry, *despite a contrary opinion within the industry*. Thirdly, relatively high capacity utilization is not reflected in terms of good financial performance in the case of large foundries, indicating price rather than volume restraints on profitability (*ibid.*). This reflects a continuous theme in the industry, that 'normal' measures of efficiency cannot explain the distribution of production or closures.

The final issue arising from a review of market perspectives is that of competition. The industry sees the issue as one of various levels of unfair competition, internationally as well as domestically, both in founding, and in supply and customer industries. A selection of comments to a WMEDU survey contained the following:

> From a world market position, the reduced demand has caused manufacturing countries to chase the available market, again with significant (dumping) price reductions.

> Many of our customers are multinational and buy their components on a world basis.

> Many of our customers have rejected any price increases and in some cases have demanded price reductions.

> . . . . but the damaging problem . . . is the suicidal antics of our fellow foundrymen (WMEDU, 1983a: 5–7).

It is the final comment that is the most revealing, and which gives a clue to the nature of the state's intervention in the industry and the problems faced by the unions.

The Origins of State Intervention and Tripartism

The long maturing crisis in the industry attracted the attention of the government as its poor performance began to be perceived as an important restraint on growth. The institution of the Ferrous Foundries Assistance Scheme and the FEDC in 1975 began a crucial period. For, as argued below, it was precisely government intervention to modernize the industry which shaped the crisis of the early 1980s, and led to the parallel development of the rationalization schemes in the steel castings sector.

The character of the state's intervention over the period is complex, reflecting not only the peculiarities of the industry but also differing policy options and political aims. The Assistance Schemes were the responsibility of the Department of Trade and Industry (DTI), and represented a particular mode of intervention which, while supported by the unions, had no institutional links with the union movement. The experience of the old Industrial Reorganization Corporation and the Industrial Development

Agency had already alerted the unions to the possibility of such intervention being monopolized by Whitehall, or captured by individual companies or groups of companies. It was for this reason, amongst others, that
the TUC had resolved to revitalize tripartism on a sectoral basis, as an
adjunct to government assistance to industry.

The establishment of the FEDC was thus of particular importance to the
unions, insofar as it represented a part of the 'new industrial strategy'
launched at Chequers that year. The disjuncture between the government
department with the decision-making power over financial assistance and
the sectoral planning body still existed, but the aim of the unions was to
use tripartism to bridge this gap, and to do so in a way that would include
the employers directly, and in a publicly visible body. This willingness to
co-operate with the employers was bolstered by the perception that many
of the industry's problems were externally generated, and that a 'producers' alliance' could form the basis of generally positive outcomes from
negotiated change. The reality of the experience was somewhat different,
as will be shown below, but it is important to remember that these two
forms of intervention ran alongside each other, without necessarily com-
plementing each other, throughout this period. This necessarily makes the
chronology of development rather complex.

The return of the conservative government in 1979 radically changed the
operation of industrial policy. The Tories blamed many of the industry's
problems on the interventionism of their predecessors and, while not
abandoning the FEDC, distanced themselves from tripartite solutions,
preferring that the industry develop according to the dictates of the
market. Since the market proved wholly unable to produce the restructuring that the industry required, new forms of active intervention were
canvassed, which led to the government-sponsored Lazard rationalization
scheme in steel castings. This scheme is dealt with in a separate section, but
neither its inception nor its operation can be properly understood outside
the general context of state activity in the sector.

In taking responsibility for a planning role, government found no centralized sectoral organization capable of representing the industry as a
whole, nor any organization which had a coherent strategy for the sertor's
development. The Assistance Schemes were certainly the product in large
part of union political pressure, but their findings expressed alarm at the
symptoms of decline rather than any plan for the sector. This failure was
even more marked in the case of the employers, as might be expected in
such a fragmented industry. The results of this disorganization were very
important in shaping the outcomes of the Assistance Schemes, and the
nature of later developments. For this reason, the schemes must be looked
at before the general work of the FEDC, as they highlight many of its
underlying problems. In doing so, it is also necessary to outline the
organizations that took part in the process.

As has been said, the organization of capital in the sector is highly fragmented. The principal divisions are by line of metal cast. The Light Metal Founders Association represents founders in aluminium and its alloys. The Association of Brass and Bronze Foundries covers copper alloys. Both are small organizations, and their business is handled by the same firm of Birmingham accountants. Zinc diecasting has its own group (ZADCA), and investment casting, principally in steel and aluminium alloys, are represented by the British Investment Casters Association.

The British Foundries Association (formerly the Council of Foundry Associations) brings together most, but by no means all, of the iron founders, and a separate structure, the British Castings Industry Research Association (BCIRA), with 700 members, organizes technical research. Steel casting is the best organized sector. The Steel Castings Research and Trade Association (SCRATA) combines trade and technical matters. A proposed merger between SCRATA and BCIRA fell through in 1984, largely because of SCRATA's unwillingness to become submerged as a minority sector, and partly because of doubts about BCIRA's effectiveness (*Financial Times*, 14 March 1984). The result has been (with the exception of SCRATA) a very low level of industry organization, with many small foundries (often owner managed) not participating, while the larger concerns, whether independent or part of larger groups, also go their own way.

None of these organizations dealt with industrial relations matters (this being the province of the Engineering Employers Federation, EEF) and had little experience or skill in organizing an industrial viewpoint, either in relation to unions or to government. Consequently, the industrial interest had tended to be dominated by leading members of the larger firms (especially Birmid Qualcast) except where an external body, such as the NEDC or the WMEDU, had actively canvassed the industry. This was to be a continuing problem as the intractable nature of overcapacity in the industry reached a critical level in the early 1980s, and led to the formation of the Major Castings Manufacturers Association in February 1983. This grouping, representing 80–85 per cent of independent producers of automotive castings, was formed with the aims of fostering a major rationalization in the iron sector, along the lines of the Lazard experiment in steel and steel alloys (*Financial Times*, 21 February 1983).

By contrast, the trade unions were, potentially at least, in a better position to view the industry as a whole. The dominant union was the AUEW(F), which has since merged with the new Amalgamated Engineering Union (AEU). Although the union recruits all grades, its craft history led to a concentration in the skilled areas and, as in motors and shipbuilding generally, there is also widespread membership in the Transport and General Workers Union (T&GWU) and the General Municipal and Boilermakers Trade Union (GMB). Additionally, there is a small craft membership in the Pattern Makers and Associated Metal Workers Union, and in the National Union of Domestic Appliance and General Metal

Workers Union (NUDA&GMW). While multi-unionism has caused some problems at plant level (NEDO, 1977a: 23) inter-union relations are generally good. More importantly, all the unions involved are organized within the Confederation of Shipbuilding and Engineering Unions (CSEU), and the three largest have always been favourably disposed towards tripartite planning.

However, the collective bargaining structure in the industry is not geared to an easy expansion of functions which could integrate tripartite planning procedures. Since the majority of the firms are affiliates of the EEF, basic wages and conditions are settled at national level, but effective earnings are settled by local, primarily plant, bargaining. In principle, non-federated firms bargain independently at company level, but they effectively follow closely national agreements and district trends. This has had the usual effect on generally small and scattered workforces of encouraging plant activity, but of discouraging combine or other inter-plant/firm organization, outside major engineering firms with their own foundry departments, such as Ford.

Outside the engineering and light castings negotiations (which account for 85 per cent of the AUEW(F)'s membership), foundry workers are covered in nine other negotiating structures, including the Ford, BL and BSC National Councils, the Shipbuilding Trades Joint Council, and the Railway Shopmen's National Council. It is in these structures where, ironically, founding is very much a subsidiary concern, that foundry unions are best placed to engage management in the kinds of discussion which union advocates of devolved planning see as the pattern for tripartism. It is also the case that decisions within these sectors have been of vital importance to the mass of foundry workers outside them.

For foundry trades unionism, therefore, the CSEU should provide a major forum for the exchange of information and the formulation of policy in respect of many of the major producers in, and customers of the sector. While the CSEU is mainly concerned with colletive bargaining issues, it is important enough for its conference resolutions to be taken seriously by government, and has regular contact with ministers, sometimes in conjunction with the EEF or Trade Associations.

Nevertheless, it is true to say that the potential of the CSEU, *even within the limited terms of its association with government departments and employers,* has been by no means fully exploited. It is the only existing trade union body in which a fragmented and largely intermediate industry such as founding could develop a coherent planning outlook, yet it played only the smallest role in the rationalization in steel castings when precisely such an overview was required.

The consequences of these failings became evident very early on in the implementation of the Assistance Schemes. The schemes originated in complaints of shortages from customer industries following the 'Barber Boom' when 'in the great dash for growth [sic] Perkins Engines for one complained that it had not been able to obtain the castings it needed from British suppliers' *Management Today*, December 1981: 35). This problem had been discussed in the NEDC in May 1975, prior to the formation of the FEDC. The Industrial Review to 1977 pointed out that 'foundries are increasingly facing problems in financing necessary re-equipment for modernising capacity, for expanding it where necessary and for improving pollution control and working conditions' (cited in *Foundry Worker*, May, 1975).

In his Budget speech that year Chancellor Healey promoted the scheme as being modelled on the Wool Textile Scheme intended to 'assist certain industries *as a whole* (my emphasis) to modernise and rationalise'. The ferrous foundries sector was advanced as 'a good example of basic industry which it is essential to modernise' (*Hansard*, 15 April 1975: 292). Details published in the autumn earmarked £80 million for the scheme, making it the largest of its type in both financial terms and in the number of projects assisted. By the time the scheme closed 404 applications had been approved, 391 of them in iron and steel. Total government assistance offered equalled £75.87 million, and total capital outlay planned equalled £345.4 million (AUEW(F), April, 1979).

The implementation of the scheme was, however, far from smooth, and its effects remain controversial. Even in 1978, DTI vetting and the onset of recession had led to 44 rejections and 65 withdrawals. Since foundries had to find 80 per cent of investment themselves, the implementation rate fell still further, so that by 1981 only £38.5 million had been paid out under the scheme (*Management Today*, December 1981: 35, 36). For those firms making investments, debt burdens and recession accelerated retrenchment. Some large firms (like Birmid Qualcast, which received £3 million in grant) had already decided to cut capacity. By 1980, several participants in the major iron founding area of the West Midlands were also prominent in the closures and redundancies listings, including Mason and Burns (£59,000 grant), John Harper and Co. (£1,293,000), RMI (£719,000) and Henley Foundries (£232,500) (WMEDU, 1982: 53). No comparable information is available for the Industry as a whole, but representatives of employers' associations, the WMEDU, and the Secretariat of the FEDC all agree that this pattern was widespread.

A major factor was undoubtedly the recession in customer industries, but the scheme also suffered from an ambiguity of purpose and, in the early days at least, an *ad hoc*, unplanned application. Since the scheme and the FEDC emerged at approximately the same time, the committee had little direct influence in the first year which saw 225 applications. By 1977 it was warning that the scheme had no clear rationale in respect of future markets or related problems, such as imports of plant and equipment. The FEDC calculated that, had all the proposed schemes gone through, capacity would have expanded by up to 25 per cent, and this led to some hardening of attitudes at the DTI (*Management Today*, December 1981: 36). Even so, in 1979 the AUEW(F) was beginning to worry about the combination of expanding capacity with declining demand (AUEW(F), April, 1979). In addition the DTI, conscious of increasing controls on pollution planned for the 1980s allowed up to 20 per cent of planned expenditure on environmental improvements (*Management Today*, December 1981: 37).

While many foundry companies welcomed the scheme in 1975, its effects have since loomed large in explanations of the particularly intransigent problems of the current period, especially in relation to price-cutting and stubborn overcapacity. The secretary of the British Foundries Association (BFA) complained in 1983 that the assistance scheme had not been founded on demonstrable undercapacity in 1974–5, a view shared by BQ's manageing director who believes that bottlenecks were the result of poor industrial structure rather than undercapacity (WMEDU, 1983b: 3). The scheme has also been criticized for its bias towards medium and large plants: 'From the days of the Foundry Aid Scheme . . . the bias of any available help is to medium and large companies . . . I find this totally unacceptable' (WMEDU, 1983a: 53).

These assessments of the scheme, of course, are not reconcilable unless account is taken of the relation between physical capacity, the quality of investments and assumptions, often unspoken, about the structure of the industry and its markets. In 1977, despite worries over the generation of excess capacity, Mr Corfield, then FEDC chairman, told the Industrial Strategy Conference that the scheme 'amounted to only £2,500 per head in a badly under-invested industry'. It was his view that double this figure was required (NEDO, 1977b: 17). Current NEDO assessments of the period emphasize very real difficulties in defining the problem to be tackled, and the relative success achieved in the steel castings sector with its more effective (if incestuous) organization. These problems were to be at the heart of the work of the FEDC for 10 years, and dominated the rationalization debates in the early 1980s.

Nevertheless, after the return of the Conservative government in 1979, it became accepted wisdom for critics of planning and state intervention that 'to some extent the problem (of overcapacity) was exacerbated by the subsidies and schemes of the last Labour government' (*Hansard*, 9 November 81: 295). In this athmosphere, and given the nature of industrial contraction noted above, rationalization 'by market forces' became the order of the day. However, the rapid contraction of the industry did not achieve much that was rational, even in tems of a free market approach. The work of the FEDC was later to underline these failures. At issue is the basis of rationalization, and its connection with improved technical and commercial performance. For the unions, however, the problem remained one of finding an independent standpoint from which to develop a policy, and to overcome its reliance on agencies which do not share their aspirations.

As long as there is a lack of such a policy, and active involvement by the workers themselves, even the most sophisticated and inclusive of unions are liable to find their sectoral activities largely shaped by the nature of the firms involved. This may have diverse results. An active and sophisticated employers' organization may provide a spur to the development of union policy. On the other hand, fragmentation and a lack of sophistication, as in the case of foundries, may contribute to a more limited union perspective. As a result, tripartite work has been even more noticeably the province of full-time officials, rather than activists, than in some other sectors, and the experience of tripartism was a major handicap in responding to the pressure for rationalization.

The Operation of Tripartism

As its inception, the FEDC was given an ambitious brief. In addition to the problems of modernization, health, safety and environmental issues, and identifying possible areas of expansion, the committee was

> to be concerned with problems of the longer term, e.g. ways of overcoming manpower shortages; energy conservation, the scope for widening the geographical limits of markets . . . mitigating the cycle; the achievement of a healthy balance between independent and 'tied' foundry operations . . . (AUEW(F), 1975: 13).

This wide remit was particularly welcomed by the unions. For the following three years their inputs emphasized the need to eradicate the 'depressingly Dickensian caricature of what modern industry could and should be' (AUEW(F), 1976). Their hopes were concentrated on achieving a high productivity, high wage industry, which would improve conditions sufficiently to end recruitment difficulties, especially of skilled labour (AUEW(F), 1977, 1978). In the early days it was equally welcomed by foundry firms, at least insofar as the FEDC was involved in promoting and vetting assistance scheme applications. In fact, however, it quickly became apparent that, outside assistance schemes, there were to be difficult problems both in terms of the general development of the industry, and on how the FEDC could develop its remit.

A fundamental problem of representation stemmed from the fragmentation of the industry on the employers' side. FEDC membership was arranged on the basis of the trade organizations, usually office bearers plus one or two foundrymen. Given the plethora of organizations involved, this led to some odd results. At one stage, a product group with only two producers had a seat, other larger groups having none. Quite apart from such obvious anomalies, this represented a more deep-seated problem, since the diversity of products and end users made it difficult for the committee' secretariat to get a clear view, and to represent founding as an industry. One result was that the secretariat sometimes developed their own analyses of the sector distinct from those of employers and unions.

After 1982, the secretariat of the FEDC attempted to use this diversity consciously to construct a more coherent representation. With some success they attempted to reduce direct participation by trade association 'bureaucrats' in favour of direct representation from leading foundry companies. The extent of the changes should not be exaggerated. Even in 1981 the committee included directors and chief executives from Butlers and Duport (to be founder memebers of the Association of Major Castings Manufacturers, AMCM) and F.H. Lloyds and the Weir Group, who were to become major supporters of the Lazard scheme. Nevertheless, the role of the FEDC was substantially about precisely this fostering of interest organization in the industry, the lack of which had been (and in many respects still is) characteristic of the industry.

The FEDC has not been noticeably successful in promoting its work through action at company and plant level. In the early days of the assistance schemes the AUEW(F)'s chairman and general secretary were involved in vetting applications and were insistent that 'where grants are approved we will expect our members to be informed and also to be concerned in seeing that all advantage be taken in using the grants in the best possible way' (AUEW(F), 1977:8).

The GMB also had a representative on the vetting committee. In general, the unions claimed that the process showed their commitment to modernization, but complaints from employers distinguished between official and shopfloor attitudes: 'management had made the scheme a success and there had been full support from the trade unions. The problem was to get the commitment from the shopfloor to ensure economic levels of operation of equipment' (NEDO, 1977b: 4). This divergence of experience between union office and the workplace was to become more apparent as rationalization proceeded.

The FEDC itself, however, also suffered from divergent views of the basic problems to be faced, and this divergence was most noticeable between management and the unions on the one hand, and the secretariat on the other. The major perceived problems for foundrymen were, as noted earlier, in terms of costs, especially of energy and raw materials, monopoly exploitation by suppliers and customers, external costs in relation to rates, health and safety, and pollution control, as well as import penetration. Several of these concerns were shared by the unions, especially in the case of imports, but they also emphasized the impact of government policy on the levels of demand and the impact of technical change.

From the point of view of the FEDC secretariat, however, these problems were seen as secondary to questions of quality and design, industrial structure and competitiveness (as opposed to 'rationalization' through closures). The result was sometimes the adoption of exercises more aimed at finding areas the parties were willing to discuss, than in getting to grips with major problems. The secretariat's initiative on quality circles is a case in point. Initially enthusiastically received, the aim was to cover problems of quality and supply, and to encourage employee participation. In retrospect, the FEDC secretaries question the relevance of this initiative, and point to its collapse after six months. Its major result, from the viewpoint of the secretariat at least, was to hold the FEDC together during a bleak period when both sides were reluctant to tackle the major issues of rationalization.

The evolution of the FEDC went through three phases: first, enthusiasm arising from the assistance schemes; then, the onset of pessimism in the later 1970s; and, finally, disarray from 1981 when the pressures for rationalization became overwhelming. This progress is amply illustrated in the attention paid to the FEDC at AUEW (F) conferences and annual reports.

From 1975 to 1978 the emphasis was on the potential of the assistance schemes to create well-paid, safe and healthy jobs. There is a marked emphasis on the theme of modernization in all its aspects, but with particular attention paid to the competitive disadvantages of poor investments. The 1978 report especially noted that 'it is obviously not possible for British employees working in antiquated surroundings with outdated machinery to match the output of foreign employees working with new, more highly productive equipment' (AUEW (F), 1978: 16). At the same time, however, the seeds of future doubts are revealed in the difficulties foreseen in technological change: 'while recognising that modernisation . . . is . . . long overdue, it has to be recognised that new investment in all industries may in the long run reduce employment opportunities . . . ' (*ibid.*).

The report of the Annual Delegate Meeting (ADM) of 1979 opened the period of pessimism proper. The long-term decline of the industry was recognized, and the failure of the assistance schemes to reverse it admitted. After reviewing trends in market contraction and materials substitution the reports warned of accelerating closures: 'it is clear that if further rationalization and closures . . . are to be avoided there must be either a revival in established markets or the development of new markets' (AUEW (F), 1979: 12). Prospects for either were considered bleak in the face of overcapacity in Europe, and a perceived threat from the Pacific area.

The 1980 report, the first since the Conservative election victory, was equally pessimistic. Noting the uncertainty of the future of the NEDC, and the TUC (and the AUEW (F)'s) determination to continue tripartism for as long as possible, the General Secretary of the AUEW (F) described 'the harsh reality . . . reflected in the continually reducing membership figures' as the result, primarily, of an international collapse of demand and consequent overcapacity. The greatest problem was identified in the steel castings sector and 'within NEDO the question of how to safeguard the industry and maintain the maximum number of jobs is being actively

considered' (AUEW (F), 1980: 5). Decoded, this amounted to an acceptance of the pressures for a radical rationalization programme that were to
mature in the Lazard scheme.

There were various tensions in the union's perception of the issues at
stake. In the first place, the previous keynote of modernization is repeated
in a further reference to the assistance schemes which resulted in 'many
foundries [having] installed new plant and improved their layout and
working conditions'. At the same time concern was being expressed about
trends towards 'deskilling', particularly where 'floor moulding has largely
given way to machines, and even machines are being overtaken by robots'
(*ibid.*).

In fact the nature and pace of technical change in the industry is far from
clear. Harvey argues that, while capital expenditure as a percentage of
gross output from 1968 to 1979 was 'hardly dramatic' (and did not change
significantly under the assistance schemes) the increase of mechanized
floor moulding and handling was a significant factor in the declining
employment rates of operative grades (1981: 555–89). The national organ-
izer of the AUEW (F) also emphasized changes, particularly in the mass
production automotive sector, from the mid-sixties. At the extreme, he
gave the example of Glyn Welsh, contractors to Ford and BL, operating
two automated plants, with 30 workers in each, producing a tonnage which
would require 200 workers using conventional techniques. At the same
time, he had found areas where automation had so speeded the process
from pouring to breaking out that labour shortages had reappeared.

Such technical developments are extremely uneven, and the difficulties
in their general introduction can be understood if it is remembered that the
established pattern of closure eroded precisely those larger, more modern
plants, where technical development was most to be expected. It is therefore not surprising that the secretariat of the FEDC expressed a far less
dramatic view. Arguing that applications for new technologies do exist in
process routes and metallurgy, they nevertheless point to a widespread
underutilization of existing mechanized techniques, and their limited applications in the jobbing sector due to small batch or one-off production.
Contrary to the union's view, the secretariat feels that it is precisely
overcapacity and cut-throat price competition which inhibits the introduction of new plant and processes; that is, *that technological unemployment
will be a problem of recovery, rather than the cause of current employment
decline.*

The Case of Small Craft Foundries
Of particular interest in understanding the development of the sector, and
of the tripartite response, is the Small Craft Foundry report referred to
earlier. Published in 1979, the report resulted from twin concerns over the
closure rate of small foundries, and customer concern over future supplies
of short-run craft castings. Despite many obvious differences, the report

encapsulates many of the problems of the industry as a whole, and makes a useful introduction to the rationalization schemes which followed. Especially important was the conclusion that:

> some small craft foundries are flourishing . . . others are being prevented from achieving more than marginal productivity by the continued existence of a third group who have no long term prospect of survival and, because they do not have to provide for the future, charge very low prices. *We think that remedial action is therefore urgently necessary, both by foundries themselves and by external bodies to ensure that a sufficient number of profitable foundries survive* (emphasis added) (NEDO, 1979: 3).

The report highlighted four main problems. Firstly, that small craft foundries were highly dependent on skilled labour, and were labour-intensive 'with costs as much as 50 per cent of selling price. Yet, because of problems of profitability, difficulties in recruitment and retention of skilled labour are a continuing threat to survival' (*ibid.*: 4).

The working party responded to these problems by seeking to identify where skill dependence could be reduced by either limited mechanization, or by paying increased attention to possible flexibility of skilled labour. The steps recommended included increased incentives, but the final formulation, agreed by the unions, had obvious implications for work practices:

> We have therefore concluded that, if a foundry, with the cooperation of its skilled operatives, could take steps to reduce its skill dependency . . . and, if their skilled employees were prepared to take on semi-skilled jobs in leaner periods, foundries would be able to offer greater potential security and higher rewards (*ibid.*: 28).

The second problem, particularly severe for small craft foundries, but applicable to the industry as a whole, was the effects of the trade cycle. Foundry output tends to enter the down phase earlier than manufacturing as a whole, and over time shows a trend to secular decline. As a result, foundries have extended periods of low demand, and have less scope for restoring profitability in favourable periods. The effects are reflected in a 'cut throat competitive situation which depresses prices even in good periods' (*ibid.*: 4), a position exacerbated by the ability of customers to exploit the situation.

The major remedy proposed for small craft foundries was membership of Trade Associations, with associations providing a forum for talks aiming at increasing profitability and cost controls. Most importantly the report recommended that 'the most immediate action the Associations should take . . . is to gauge the industry's support for co-ordinating the timing of price increases, and discuss them with the Department of Trade' (*ibid.*: 8).

The third major feature of the report followed from problems associated with supplier strength, that is 'the reliance on large, dominant suppliers, for its key supplies of pig iron, coke, gasses and sand' (*ibid.*: 5). This problem cannot be tackled directly, as in the case of output prices, despite its adverse effects on cash flow and working capital. It does, however, add strength to the call for strong industry association and co-ordination with external bodies. As far as the industry as a whole is concerned, the British Castings Industry Research Association (BCIRA) played a strong role in the lobby for government subsidies on energy costs, and achieved a temporary subsidy on coking coal.

The fourth element of the report focuses on the state, and in particular on the issues of health, safety and the environment. The report claimed that 'expenditure to meet the requirements of Health and Safety and environmental regulations will amount to about 75 per cent of necessary future investment; we consider that no other small business sector is facing such a prospect' (*ibid.*: 4).

The initial response of the working group to this problem included pleas for financial assistance in line with existing French and German schemes (*ibid.*: 85), but is remarkable for the emphasis of its recommendation that 'foundrymen query any instructions from the Health and Safety Inspectorate, the Alkali Inspectorate or the Local Authority which they think inappropriate' (*ibid.*: 6). Particular attention is paid to the role of local authority control of air pollution since foundries 'have shown how vulnerable they are to demands . . . to invest in *unnecessarily elaborate . . . control equipment*' (emphasis added). The report therefore recommended DOE guidelines to local authorities to 'encourage them to be more reasonable' (*ibid.*: 8). The agreement with these positions on the part of the unions strangely contradicts their aspiration for the safest and healthiest possible working environment which tripartism was to have encouraged.

It is in the general tenor of the report, however, that the insistent problems of the sector are to be found, problems which may be modified in the larger firms, but which nonetheless characterize the sector as a whole. The central thrust is in fact towards radical changes in work practices, an early and controlled elimination of part of the sector, the need for a closer industrial organization, verging on a cartel, to ensure this, and the necessity for state action to sponsor the entire scheme.

The latter point is reinforced by the discussion on the possibilities of mergers. Recognizing the potential difficulties in the early elimination of foundries that 'will have to close in five to ten years', even with strengthened trade associations, the report pointed to the need for foundrymen to discuss possibilities 'with an honest broker, who might provide the necessary catalyst. The Department of Industry Small Firms Service has a scheme to help finance feasibility studies' (*ibid.*: 7).

The report was compiled and published before the optimism of the assistance schemes had totally evaporated. Union acquiescence, particularly in the matter of mergers and closures, could still be reconciled with the object of stabilizing employment, though their stand on work practices and health and safety could undermine the interests of the workers in the industry. Employers' special pleading over external costs, especially those generated by the government and local authorities, is readily understandable. Accord over the forms of rationalization, however, was more of a problem, and became increasingly so as the industry deteriorated further. At the turn of the decade, the clearest view appeared to emerge from the secretariat which was unable to dominate thinking, not only because of internal stresses within the industry and the FEDC, but also because of the neo-liberal approach to industrial policy being adopted by the Conservative government. The interplay of these forces forms the background to the attempts in the steel castings sector to organize a new form of co-ordinated retrenchment, involving government subsidies for a scheme based on purely market criteria, and organized by the merchant bank Lazard Frères.

Lazard Frères and the Rationalization of Steel Castings

The persistence of overcapacity, and a growing conviction on the part of the larger employers that the reluctance of the less efficient firms to quit competition was a drag on the market, led to increased interest in an organized rationalization. The FEDC secretariat was particularly insistent on the need for such a programme since evidence indicated that factors other than productive efficiency (financial strength, the stubbornness of small family firms) meant that the pattern of closures was failing to promote a new industrial structure capable of serving future demand.

While such a scheme was proving impossible in the ferrous sector generally, new developments in steel castings emerged which cut across tripartite perspectives and opened a new experiment in a quasimarket-based rationalization. With government support, Lazard sponsored a plan aimed at reducing capacity by 25 per cent. This scheme not only side-stepped attempts to form a tripartite consensus, it also challenged the rationale of the FEDC's attempt at a planned approach intergrated with an analysis of prospects in customer sectors. In short it represented a new form of organization by finance capital and the government of market forces.

Ironically, the seeds of the Lazard scheme lay in the experience of the 1975 assistance programme itself. As previously noted, the random nature of that assistance, in conjunction with an accelerated decline in demand, increased capacity while at the same time placing further strains on firms' finances. The steel castings sector, being smaller and dominated by relatively large producers, and with the most efficient trade organization in the industry, reacted far earlier than the iron sector.

Steel castings reacted by cancelling or scaling down investments proposed under the aid scheme. SCRATA produced its own assessment of future demand, and broadly agreed with the FEDC secretariat that contraction was more likely than growth. The Association raised its concerns with the FEDC, not least because there seemed no other arena where the mechanisms of any rationalization could be worked out. In addition SCRATA was, at that stage, anxious to secure trade union consent to any measures adopted.

This desire to secure union consent, however, may well have reflected little more than the habit of tripartite work. Certainly events were to show that SCRATA, and the employers generally, were unwilling to engage in much by way of consultation, let alone bargaining, once finance became available outside the tripartite framework. In fact, the scheme became, simultaneously, an exercise in the political isolation of the unions.

Even so, there was one set of factors which limited the employers' disengagement from tripartism. The market-led, finance-based scheme, which was developed by Lazard, always held the danger of misallocating resources because it could not directly include non-monetary factors such as capital structure, skills, training and productivity. These failings were to push the employers back towards the FEDC at a later stage, but this was the result of the outlook and analysis developed by the secretariat rather than a felt need to involve the unions. The cycle of events that follow in no way invalidates the proposal that the unions were marginalized within as well as outside the tripartite structures.

In September 1979 SCRATA pressed the case for rationalization at the FEDC. Agreement on the need for a scheme was reached in principle, but it was accepted by all the parties that the industry could not act without additional funding from the government. The government expressed sympathy for the projected rationalization, but refused to contemplate financial assistance, ruling out the use of £20 million left over from the original assistance scheme.

At that stage, the tripartite institutions seemed to have run out of steam, to be replaced by privately run initiatives of major companies in the sector, who approached Lazard for advice (*Investors Chronicle*, 24 July 1981). The origins of this initiative remain obscure but seem to have involved contacts between Viscount Weir of the Weir Group, and F. H. Lloyds with Ian Macgregor at BSC, which itself had a direct interest in relation to its Sheffield founding subsidiaries. The managing directors of both Weir and Lloyds were FEDC members, and had been closely involved with SCRATA'S original initiative.

Lazard organized two schemes, one for the high alloy sector, and one for the general (light and heavy) steel castings industry. The plan centred on voluntary closures of capacity to be funded over a five-year period by contributions from the remaining firms. At this stage, government assistance was confined to proposals for easing possible financial restraints. Since closers would find their highest costs at the point of closure, Finance

for Industry was considering discounting promissory notes from the openers to facilitate lump-sum compensation. At the same time, the Department of Industry agreed to certify the scheme under the 1970 Corporation
Taxes Act to allow openers' payments to be tax deductable (*Financial
Times*, 23 November 1981).

A central feature of the Lazard plan was a strict enforcement of capacity
cuts. 'Melting furnaces, and ancillary equipment will normally have to be
destroyed, and companies . . . will be required not to engage in any
British enterprises to make or distribute steel castings for five years'
(*ibid.*). The formulation was striking for two reasons. Firstly, while it
obviously was a cartel arrangement of sorts, it differed in that it proposed
no new form of organized production, price controls, or limitations of
competition amongst openers. Secondly, the commitment to the physical
destruction of plant (the most immediate and reliable form of the devalorization of capital) was clearly an attempt to give market forces a prod
without actively considering a plan for capacity in relation to the wider
market, or internal balances of employment, skills, capacity type, etc.

Such a stringent reliance on the market was unacceptable to many
foundrymen who were concerned that the transition period might leave the
industry unable to meet customers' requirements. Accordingly agreements
in both sectors included terms intended to ensure continuity of supply and
an ordered transfer of work. Lazard claimed:

> It is not expected that the availability of castings will be affected and
> the reconstruction agreement . . . incorporates vigorous provisions
> to ensure that there will be as little disturbance as possible. The
> provisions include . . . a contractual obligation on closing foundries
> to give information on products and customers to the participating
> foundries remaining open (Lazard Frères, 1982).

Although the high alloy sector is small (the 16 companies finally participating accounting for some 95 per cent of output), it was not easy to secure
companies' compliance. In October 1981, only 10 had agreed and the
scheme threatened to founder 'as some 40 per cent of the industry is
standing off hoping for a free ride from participants' (*Financial Times*, 23
November 1981). The same problems subsequently delayed the general
sector scheme until 1983. Attempts at agreement repeatedly suffered from
the generally low levels of interest organization in the industry, and Lazard
found itself increasingly courting the major companies in private. This lack
of organization was apparent at a steel castings conference called in 1981
by Lazard and SCRATA where, as one participant put it, 'we were lucky
people did not come to blows there. Some people merely turmed up to see
if they could get their competitors to admit they would close down to use as
ammunition with customers' (*Engineer,*15 October 1981).

One result of this reaction was government agreement in December of
that year to make grants under the 1972 Industry Act, and to agree that no

TABLE 2.3
Foundries Closing under the Hazard Rationalization Scheme, 1981

Holding Company	Foundry	Government Compensation(£s)
Spencer & Halstead Ltd	Jonas Webb Ltd	198,750
Thomas Carling & Co. Ltd	T. C. & Co. Ltd	63,600
GKN plc	Sheepbridge Alloy Castings Ltd	202,725
Law & Bonar plc	Bonar Langley Alloys Ltd	294,150
Wellman Eng. Corp.	Wellman Alloys Ltd	n/a

Source: *British Business*, July, 1982

claw-backs would be made on the basis of previous funds received under the 1975 assistance scheme. This turn-around was partly due to increased government interest in the possibilities of this type of scheme, and the arrival at the Department of Industry of Patrick Jenkins, who was markedly less enthusiastic about neo-liberal orthodoxy than his predecessor.

Nevertheless, even this concession nearly proved insufficient. Some major companies threatened to stand aloof, or engaged in hard bargaining for their own corporate interests. Only sizeable government funding, and persistent pressure from Lazard persuaded these companies finally to back the scheme. The agreement signed in March 1982 covered 16 foundries, accounting for 95 per cent of total output. Closures accounted for 22 per cent of output, and job losses were 400 out of the 1,500 in the sector (*Financial Times*, 24 March 1982). Closers are shown in table 2.3 with amounts received from section 8 funding under the 1972 Industry Act.

Both schemes for steel castings were characterized by intense secrecy and arm-twisting, interspersed with bouts of speculation in the trade and financial press. An early report in the *Engineer* quotes the managing director of Darwin Alloy Castings (eventually included as an opener in the high alloy scheme) as saying, 'the first I heard was when the Works Manager told me our shop stewards were upset at seeing a clipping in a local paper linking our name with the plan' (*Engineer*, 30 July 1981: 20). the *Engineer* emphasized a point that was to remain broadly true for three years when it concluded that, 'only merchant bankers Lazard and a select few giants in the industry including BSC, the Weir Group and F. H. Lloyds were party to discussions which led to the proposals' (*ibid.*).

Nevertheless, towards the end of 1982, Lazard began to worry that accelerating decline was overtaking their original plans. Forced to end speculation, and to finalize negotiations that had been 'a tightrope walk all the way through' (*Engineer*, 17 February 1983), Lazard published a dead-

TABLE 2.4
Steel Casters Rationalization Scheme

Openers	Closers and Compensation (£)	
BSC Craigneuk Medium Foundry	H. Broadbent & Sons Ltd	513,691
Sheffield Forgemasters Heavy Foundry (River Don)	Sheffield Forgemasters Railway Foundry	318,621
Brockhouse Castings	BSC Craigneuk Light Foundry	373,641
Lake & Elliot Founders & Engineers	F.H. Lloyd (Wednesbury)	2,769,030
Lloyds Burton	Holcroft Castings	331,627
Parker Foundry	Head Wrightson Steelcast	1,111,911
National Steel Foundry	Ryder Brothers	113,041
North British Steel – Armadale Foundry	Triangle Valve Co. Ltd	177,565
North British Steel – Bathgate Foundry	Total	5,709,127
Weir Group – Catton & Co		
Weir Group – Jopling & Sons		
Weir Group – Holbrook Precision Castings		

Sources: British Business, May 1983, *Engineer*, 17 February 1983

line for the scheme in January 1983, threatening to abandon the whole plan if arrangements were not finalized by the end of the month.

This urgency was underlined by the deepening crisis. At the inception of serious discussions in 1981, closure targets had been around 15 per cent capacity, with estimated government aid of £2.5 million. By the start of 1983, Lazard was looking for a 30 per cent reduction, and government had increased its incentives to £6 million, at the cost of an estimated 2,200 jobs. In addition to the physical destruction of capacity requirements of the high alloy scheme, 'closers' now had to agree to keep out of founding for 10 years (*Engineer,* 17 February 1983). When the scheme was finalized in February, 12 companies out of 35 allocated to the sector had agreed to close 10 of their 22 foundries. Because of last-minute withdrawals by four companies, and attempts by three others to set up a separate scheme for high volume machine moulded castings, capacity reduction was 25 per cent rather than the 30 per cent envisaged (*Financial Times,* 12 February 1983).

The list of openers and closers included some companies in both categories, with the result that they were committed to a levy which would help compensate themselves, as well as receiving government compensation. F. H. Lloyd, an initiator of the scheme, and the largest single producer, closed its Wednesbury foundry on which they had expected to make a £1 million loss in 1983. They received £2,769,030 compensation from the government, the largest single payment. Sheffield Forgemasters railway foundry had been taken over by BSC in the certain knowledge that it was in serious trouble. By keeping the River Don works open, and closing the railway foundry BSC received £318,621. The corporation received an additional £373,641 for closing one of its Craigneuk plants, and keeping the other open. Total payments (excluding Wolsingham Steel which held out for special treatment) amounted to £5,709,127. BSC Craigneuk had received £1,228,800 in grants under the 1975 assistance schemes (AEUW (F), 1979: 16, and 1983a).

Union Responses

The minimal role played by the FEDC in Lazard's highly secretive set of proposals put the unions at a major disadvantage. Reports of the scheme began to appear in the press from July 1981, leading to demands from the AUEW (F) that the FEDC discuss the issue. A meeting was arranged in September with representatives from Lazard and SCRATA, and a further meeting (not attended by the AUEW (F)) was also held to discuss the scheme in general terms (AUEW (F), 1981). The unions proved unable to use political pressure effectively to escape from a dependence on rumour, an experience that was to recur throughout the period.

As a result, it is hardly surprising that the unions were virtually unable to act in the case of the high alloys scheme, but this isolation and dependence on rumour continued to characterize the unions' position throughout. In February 1982 the Foundry Unions Committee of the CSEU wrote to both the FEDC and Lazard 'to request an up-to-date report on the latter's activities'. No reply was received from Lazard until the CSEU received a copy of their press release on 29 March (CSEU, 1982). The reply from the FEDC emphasized this isolation and the extremely limited role played by the secretariat. After pointing out that the issues had been discussed at the FEDC and that two meetings had been arranged with representatives from Lazard, it continued 'I think it can be taken for granted that all of us . . . are equally anxious to see the end of this particular development, and to have all possible information available to us' (FEDC, 1982).

However, by July 1982 the National Council of the AUEW (F) was again reporting investigations into 'rumours that the rationalization plan for the steel castings sector is about to re-emerge'. The Executive Council sought meetings with McGregor, the Department of Industry, and the FEDC to get further details, and followed that with an appeal for information from members since, 'this seems to be only way we are going to be able to tackle this problem, rather than be faced with the same situation as took place recently with the High Alloy sector' (*Foundry Worker*, July 1982:110).

As the rumours took on a more concrete form, a delegate meeting took place in Sheffield on 25 August, with 58 delegates from the major unions, and a resolution calling for complete opposition to the scheme was carried unanimously. The conference was not a success, however, and never led to an active response to the scheme. As the meeting was called before the lists of 'openers' and 'closers' was published, in order to prevent a split, the opposition to the scheme was described by one official as 'being against Lazard like we are all against sin – hot air – no strength or strategy' (AUEW (F), 1984).

The next step taken by the unions had little more prospect of success. A resolution was submitted to the TUC, calling for a campaign 'to prevent the demise of this vital basic industry' (TUC, 1982), but did not specifically mention the Lazard scheme, nor propose a strategy for such a campaign. In the event, the AUEW (F) withdrew the motion on the request of Congress House because of pressure of business, but on the understanding that the General Council would consider the matter. Accordingly, a report was forwarded to Congress House, in which the AUEW (F) reiterated the part it had played in the FEDC and the role of the assistance schemes. It then added a history of the industry's decline, and outlined the Lazard proposals, concluding that,

> it should be noted that while representatives of the steel castings employers sit on NEDO, despite the tripartite nature of that body, the Foundries EDC have not been involved as such. Neither have the trade unions which serve on the Foundries EDC (AUEW (F), 1982).

The report emphasized the problem of accountability raised by the use of Industry Act funds in the rationalization scheme, the threat posed to the rest of the ferrous sector, and fears that a 'rationalized' industry would be unable to meet future demand, and would fall prey to import substitution. As might be expected, given both the isolation of the unions, and their inability to propose or organize a campaign, the TUC's response was not radical, promising consultation with the CSEU, and proposing to write to the Secretary of State for Industry expressing their concern.

In a final attempt to influence at least the implementation of the scheme, the CSEU agreed to official support for any foundry workers resisting closures, and called another delegate conference for companies involved on March 1983, in Sheffield. The report from the Foundry Unions Committee to the Executive Council of the CSEU following the conference emphasized the difficulties of organizing resistance in an industry facing a high closure rate, the isolation of officials from the shopfloor, and union concern about the broader implications of such forms of rationalization scheme. It is these implications which must now be considered.

At every new development, the ineffectiveness of the unions' reliance on its membership of the FEDC was confirmed. During and after these events, the FEDC played a role similar to that during the high alloy scheme, in that it acted primarily as a highly ineffective contact point for union information-gathering. The secretariat itself was little involved, and limited itself to monitoring published proposals for any obvious distortions to supply. The AUEW (F) expressed its despair over this state of affairs in a long letter to the FEDC in which it complained that it had to rely on press reports for information.

The letter, however, expressed not only opposition to the Lazard scheme, but also the ambiguities in the unions' own perception of the problem which underlay their inability to initiate a vigorous response.

> As you are aware, some two years ago the suggestion was made that there was always the possibility that restructuring of the industry might become necessary. To the best of my recollection my response was that in the event the trade union representatives would seek to get the best possible financial compensation for their members in respect of them losing their jobs.

> Despite the fact that considerable sums of money are said to be involved . . . there is no provision for enhanced redundancy payments (AUEW (F), 1983b).

Quite why firms should have been expected to part with any of their hard earned subsidies at this stage is unclear since there seemed little the unions could do to apply pressure.

The unions' investment in the 'new industrial strategy' on the other hand had already committed them in many ways to the rationalization strategy. In relying on developing relationships with the state in order to influence the forms (and compensations for) restructuring, they had ignored the possibilities of fostering an independent programme. Partly this was the result of technical deficiencies in their ability to formulate such a policy. More important, however, were the political restraints inherent in their strategy. Any restructuring which failed to challenge the ownership structure of the industry and its relations with its major customers could result only in sacrificing one section of the workforce in the interests of a section of capital. Such a radical challenge could not be mounted inside a structure which gave legitimacy to the private ownership and control of capitals in the industry, unless it were to degenerate into a *de facto* alliance with the major, and potentially dominant firms of the sector.

The Nature and Results of the Lazard Scheme
In many ways the Lazard scheme can be seen as a new form of 'industrial strategy' compatible with the economic liberalism and hostility to trade unions characteristic of the present government. Its emergence in steel castings was by no means accidental, even leaving aside Lazard's close

connection with BSC, but was closely connected with the development of the 'Phoenix' programme for a general rationalization of the steel industry (*Engineer,* 30 June 1981). Original discussions between Lazard and the Bank of England envisaged action in the whole ferrous foundry sector, drop forging, steel rerolling and wire drawing, although thus far only wire drawing has an agreed scheme (*Financial Times,* 12 February 1983). The Lazard approach appealed to both the government and those industrialists who rejected active and discriminatory state intervention, but who sought to subvert rigidities in markets which were inhibiting 'natural' rationalization. Lazard explained their outlook thus, 'It should be emphasised that the scheme . . . is entirely voluntary and that the decisions leading to participation are on commercial grounds . . . [The participators'] view is that the scheme will be less disruptive than the operation of inevitable, but slower, market forces' (Lazard Frères, 1982).

The fact is, nevertheless, that the 'voluntary' nature of the scheme hid the application of considerable muscle by market leaders. In turn this implied the possibility at least of a sufficient degree of power in the hands of these firms, even where this did not amount to a clear oligopoly, and the ability to mobilize support external to the industry, *including state support.* This was candidly admitted at the outset in 1981. 'The aim was for the big boys to put their money where their ideas are and then to drag the small boys in with them. If the reshuffling does not take place of its own accord, then the smaller companies may have to be given a prod by an outside report' (*Engineer,* 30 July 1981).

What is at issue in effect is the creation of a cartel, but a cartel with a difference. The cartel, as classically understood, involves the dual function of denying new entrants to a market, and regulating output, prices, and competition between members. Such activity has a long history in the UK, and was pursued actively in the inter-war slump, with a large degree of connivance by the state. Such cartels, however, require sophisticated organization, well-formulated rules, and feasible sanctions, and have in any event often proved extremely unstable in times of crisis. The Lazard scheme was different. It sought, by concerted action, to reduce capacity to such a level that orthodox competition might apply. As a result, its aims were limited in time, destructive rather than regulatory, and sought to combine the advantages of oligopolistic power with the virtues of the free market.

The corollary is a most important contradiction. On the one hand, capital must be capable of a degree of organization and self-discipline if the coup is to come off. On the other, different capitals cannot enter too closely into association with those with whom a renewed competitive battle is shortly envisaged. This is a very fine line to walk, and largely explains why the high alloy scheme (a very small population of firms, with a relatively high capital concentration) took a year to organize, and the general castings scheme (with a large population, and a higher proportion of small producers) took two years. It is also a major reason why a scheme for iron foundries never got off the ground. Thus while Lazard were confident that 'this new approach can be applied to sectors other than steel' (*Financial Times,* 24 January 1983), the exercise proved abortive.

A feature of the scheme, based on relative financial strength and the prospect of future unregulated competition, was its inability properly to define the relevant product and market mix that had been a central concern in preceding tripartite work. This difficulty was aggravated by the fact that the rejection of any continuing central co-ordination, which could have replaced, in part at least, the sectoral indicators of tripartism, meant that the policing of the scheme was highly haphazard.

The *Engineer* reported in February 1981, one year after the implementation of the high alloy scheme, that 'pirate foundries and poor market predictions have left many of the foundry owners . . . disgruntled at having paid out millions of pounds to be in exactly the same position they were in last year' (*Engineer,* 17 February 1983:8).

The report claimed that none of the foundries gained the expected boom in orders due to a combination of continued decline in demand, and the fact that 'people producing iron and steel castings have moved into the high alloy sector, attacking the markets those staying open had paid to protect' (*ibid.*). The most blatant case was the creation of a new company, Armalloy, when managers bought up the equipment of the closing Wellman Alloys Stourbridge plant, and reopened in new premises. To add to the confusion the company received a £70,000 loan from the WMEB, and investment capital from the National Water Board's pension fund (*ibid.*).

In the circumstances it is not surprising that the FEDC as a whole viewed the scheme with some scepticism. A discussion paper produced by the secretariat emphasized that its primary concern was with cutting capacity, but continued, 'while desirable, the attainment of these objectives will not necessarily secure the long term development of the industry' (FEDC, 1983c). The secretariat proposed instead a structured and planned scheme including non-money factors, such as skills, labour supply, the needs of specific plants, location etc., and listed seven further considerations including product range, specialized customer requirements, the trade balance and the retention of union goodwill (*ibid.*).

In various ways, then, the experience of the Lazard scheme produced something of a revival of interest amongst employers and unions in a tripartite discussion of rationalization. The AMCM again discussed rationalization measures, and the secretariat produced a paper, 'The Next Steps', which emphatically rejected a free market solution to the industry's problems (*Financial Times*, 11 January 1984). This report (billed as highly controversial, and as yet unpublished) noted the continued decline in output and employment, but argued that, 'this process of attrition has been operating for a number of years [but] has so far failed to result in a slimmed down, efficient industry' (*ibid.*). More importantly, it proposed that secu-

lar decline in demand, combined with materials substitution and innovation in technology and design, meant that even an industrial recovery
could only slow down, rather than reverse that decline, unless a substantial
rationalization took place. It concluded that 'additional capital investment
. . . will be productive only if undertaken by a minority of present
companies and if total present manufacturing capacity is reduced' (*ibid.*).
The FEDC thus returned to the basic problems of industrial structure and
overcapacity, but this return did nothing to resolve the problems which had
faced both the assistance schemes and the Lazard operation. The conclusions in respect of closures had long been the viewpoint of the AMCM,
but had also proved the stumbling block in all attempts to organize the
owners of the crucial iron sector.

At this delicate stage, the decision by Ford finally to close its Dagenham
foundry intervened, and caused the trade union side of the FEDC to block
a proposed consultants' report on European prospects for automotive
production (*Financial Times*, 20 January 1984:8). The unions objected that
while 'we are not deaf to proposals . . . we are not prepared to write a
blank cheque for anyone' (*Engineer*, 26 May 1984:8), and pointed to the
possibilities arising from a domestic redistribution of Dagenham's 70,000
tons per year output. However, the unions' ability to develop concrete
proposals on this basis was undermined by the disclosure that Ford had
been in discussion with the AMCM for the previous two years, and that 80
per cent of the work would go to Germany and Spain as part of Ford's
multinational sourcing policy. In fact, the AMCM proposals, which were
central to the developing plans of the FEDC, had *already* taken the results
of the Dagenham closures into account (*ibid.*).

Conclusions

The 'new industrial strategy' of 1975 has left behind only the institutional
traces of the unions' aspirations to link planning and collective bargaining,
but the ideas developed then remain the basis of much of the movement's
current thinking. Little political accounting has taken place, even though
today the political and economic assumptions underlying tripartism seem
much less compelling. If a real understanding of the role of labour in
restructuring is to be developed, then not only must there be a political
reassessment of the part played by the unions, but also of the nature of the
changes the industrial economy is undergoing; that is, there is a need for a
political economy which can address the issues in a coherent way.

The case of foundries throws doubt on any simple economic or institutional ideas of restructuring. The most striking feature of the industry
has been the consistent failure of the market to achieve the restructuring
which could have been expected in a period of long decline accelerating
into crisis in the 1980s. Nor have the various interventions of governments,

their agents, or outside bodies managed to resolve the problem. These failures raise some interesting questions regarding the nature of industrial restructuring generally, and in particular for the future of those industries which would be marginal in terms of regenerating profitability as a whole.

The nature of competition in a period of crisis is crucial in this respect. The ability of British industry to solve its structural problems is crucially dependent on the devalorization of capital, and this is the underlying thrust behind the wave of closures that has hit manufacturing. In many ways the neo-liberalism of the Conservative administration is reflected in this requirement, and the evident failures of previous interventionist policies to achieve it. However, a simple faith in the market, a nostalgia for the supposed competitive virility of early capitalism, is proving a chimera. Not only is there little evidence that this process can foster the positive adjustments needed, it is equally clear that competition alone will not necessarily achieve even the destructive ground work.

The foundry industry is highly fragmented, both in terms of products and capital concentration, an industry in which competitive distortion should be rather low. From the point of view of rationalization, however, it would be unrealistic to characterize it as a part of the competitive sector of the economy, as distinct from the monopoly sector. What is important is to understand the industry within the context of an economy dominated by high capital concentration and monopoly power. On the contrary, the sliding scale of dependency between tied and wholly independent foundries, and consistent cost problems associated with the weak position of the industry *vis-à-vis* monopolized suppliers and customers has resulted in marked distortions of the presumed role of competition, which has hindered rather than fostered rationalization. This reflects a deep-seated problem which relates to both capital ownership and organization. Considerable evidence has been accumulated to indicate a lack of organization, through trade associations or other forms of interest representation, which could have mitigated the intractable problems of price-cutting, and general anarchy in the sector. *In many ways it is the absence of monopolistic features, rather than otherwise, which has inhibited change.*

In the circumstances, it is wholly unsurprising to find such an important part being played by external agencies. One of the main effects of the intervention of the FEDC was an attempt to foster just such an internal organization, not only in the hope of modifying the competitive behaviour of firms, but in order to establish a sector-wide basis for a rationalization programme based on a thorough account of the market and its likely development, a process which is so often generated internally by large firms through both buying and marketing strategies. The Lazard scheme too, though eschewing the trappings of indicative planning, aimed at much the same end; a rationalization to suit the larger firms which may then have been able to develop their own internal strategies, freed from the importunities of smaller firms, more concerned with simple survival than anything else.

The state played a vital role within this process. The role of the FEDC would not have been even as developed as was in fact the case without the use of Section 8 of the Industry Act to channel substantial funds to the industry. Even the Lazard scheme, apparently a free market initiative, required state funding, and authorization under competition rules. Furthermore, even before government intervention, the Lazard scheme relied heavily on the consent of the Bank of England for discounting facilities. Lazard was only another link in a chain of political influences on the industry's development which ran from the government, through the FEDC, the Bank of England and, in some areas, local councils or economic development units. Within these relationships were subsidiary stresses, as for example the FEDC staff preference for a planned rationalization against Lazard's market-led scheme, and central versus local initiatives. At the fringes, but of some importance still, were other central and local government agencies with important roles impacting on the industry, especially in terms of health and environment issues.

The failure of the industry to overcome its problems autonomously and the interventions of the state and other agencies engaged the unions in a specific set of relationships and negotiations, but a set which, politically, owed much to the union movement's own thinking. The two main phases of the case studied here reflect very different government attitudes to the role of the state and the status of the unions, both of which presented different sets of problems for the unions in operating and developing policy. Many of these were specific to the industry, but many have a more general interest too.

The development of TUC policy from national, through sectoral, to corporate and plant planning procedures arises from many sources, but is centrally the expression of a desire to achieve a bargained management of the economy at all levels of decision-making. The necessity for such a policy is based on the inadequacy of its relations with the employers directly (that is, through collective bargaining) to protect the interests of the membership, and hence we see union reliance on a relationship with the state. Much has been written on this theme, but two important aspects of the experience in foundries may usefully be considered here. The first is the connection perceived by the unions between their collective bargaining activities and their participation in tripartite planning bodies. As far as industry-wide bargaining is available to the unions, it is conducted with the EEF which does not, and cannot, bargain over many of the issues which would be relevant to the planning process. The second, and in many ways more important, bargaining level is concentrated in the plant. The fragmentation of the union workforce is thus particularly marked, resulting in many ways from the fragmentation of the industry itself. This means that intra-company combines (let alone intra-plant organization) are non-existent, nor is it easy to see how an appropriate organizational expression could emerge within this bargaining framework which could realistically formulate a planned response to the crisis in the industry. Even were such an organization to emerge (and the Sheffield conferences mentioned earlier could indicate an initial form), the problem would still remain that the employers themselves are in no position to undertake negotiations which would include investment, manpower, training, and so on, as there exists no organization which could mediate the process.

This is stating the issue in its starkest form, but it is a conclusion borne out not only theoretically, but empirically. In the period of the assistance schemes, it is true, attempts were made (or rather claimed) to involve the shopfloor in the introduction of new plants and working methods. To conflate the inevitable bargaining over such issues with planning is, however, to succumb to an illusion. Even when the assumption still held that participation was in the context of growth, the operation of the scheme had no coherent criteria for product or technical development, and the position was infinitely worse as overcapacity came to dominate the industry's thinking. In reality, there appeared an unbridgeable gap between the activity of the union officials on the FEDC and the bargaining activity of lower level officials and lay activists. The whole experience of the Lazard scheme demonstrated quite effectively the continuing gap between collective bargaining structures and concerns, and participation in, and the political advocacy of tripartite planning.

If that were the only problem, it might be argued that adjustments in bargaining arrangements could seriously modify these difficulties. The second theme, however, raises even more serious objections to such a view of the trade union role.

The closure of Ford's Dagenham foundry highlighted the difficulties of planning an intermediate sector such as foundries, indeed the difficulties of its definition as a sector. While unions engage in discussions with employers within the industry, the agenda is being set elsewhere. No rational approach to the problems of the industry, *from the workers' viewpoint,* can be considered if it excludes this consideration. When those discussions are conducted at the level of tripartite bodies, these difficulties are in no way overcome, and in some respects are exacerbated.

The attempts, through tripartism, to construct a sectoral interest point to one way of overcoming capitalism's own inability to rationalize the sector. The success of this project, however, runs against the interests of the workers within the industry, since it would not only 'slim down' the population of firms, but give considerable impetus to the process of technological unemployment. Plainly, workers in the industry (workers everywhere) have no interest either in supporting current structures, which have resulted in such decline and social waste, yet it cannot be assumed that the interests of the surviving firms represents also the interest of the workers. This was partially recognized by the unions in their approach to rationalization, especially insofar as they argued that the process must include action by the state, at the social level, to resolve the problems of those driven out of work.

This, however, is simply to accept the logic of a particular capitalist response to the problem, and a response that has proved inadequate in all its manifestations in this case. There seems little future for a response that takes as its starting point simply the sector itself, and capital's own definition of the problems, in order to reduce a large portion of the workforce to clients of the welfare system. Rather the very problems uncovered in the last few years should be a starting point for the development of an independent approach, most especially in respect of the destructive relationships between capitals themselves.

Thus, it may be the case that a socially rational development of the industry would be possible only on the basis of planning in the major customer sectors. This is hardly likely to appeal to the employers in those sectors who have done moderately well by exploiting foundries, but it does offer a linkage which could resolve the contradiction noted earlier; that where the unions are strongest is in precisely those industries that hold the fate of foundries in their hands. Within this perspective, the only feasible union role would lie not in collective bargaining with foundry employers, but rather in discussions with the workers in the major customer industries. Furthermore, given the process of change in materials and design, not to mention product mix and development, such a relationship would have to be developed on the basis of engineering as a whole, and not the foundry sector alone, if job transference and training were really to allow for a controlled and socially rational restructuring of the industry.

Bibliography

Amalgamated Union of Engineering Workers (Foundry) (AUEW (F)). 1975. *Conference Report*. April.
—— 1976. *Conference Report*. April.
—— 1977. *Conference Report*. April.
—— 1978. *Conference Report*. April.
—— 1979. *Conference Report*. April.
—— 1980. *Conference Report*. April.
Amalgamated Union of Engineering Workers (Foundry) (AUEW (F)). 1981. Letter to the FEDC from the General Secretary. 27 November.
Amalgamated Union of Engineering Workers (Foundry) (AUEW (F)). 1982. Unpublished statement to the TUC General Council (mimeo) 13 September.
Amalgamated Union of Engineering Workers (Foundry) (AUEW (F)). 1983a. Internal memorandum (mimeo).
Amalgamated Union of Engineering Workers (Foundry) (AUEW (F)). 1983b. Letter to P. Benson, Secretary FEDC from the General Secretary. 28 February.
Amalgamated Union of Engineering Workers (Foundry) (AUEW (F)). 1984 (field notes) 22 February.
Baden Fuller. C. and R. Hill. 1984. *Industry Strategies for Alleviating Excess Capacity: The Case of the Lazard Scheme for UK Steel Castings*. London: London Business School.

Confederation of Shipbuilding and Engineering Unions (CSEU). 1982. Letters to the Foundries Union Committee, 24 and 29 March.

Engineer, various dates.

Financial Times, various dates.

Foundry Worker. AUEW(F). 1975. May.

——. 1982. July.

Foundries Economic Development Committee (FEDC). 1982. Letter from T. Kilpatrick, Chairman, to A. Ferry, General Secretary, CSEU. 18 March.

Foundry Economic Development Committee (FEDC). 1983a. 'Assessment of Future Demand. FS(83)21' (mimeo).

Foundry Economic Development Committee (FEDC). 1983b. 'Assessment of Future Capacity. FS(83)22' (mimeo).

Foundry Economic Development Committee (FEDC). 1983c. 'The Steel Castings Rationalisation Scheme – Note by the Office. FS(83)6 (mimeo).

Hansard. 1975. HC Deb. 292. 15 April.

——. 1981. HC Deb. 295. 9 November.

Investors Chronicle. 1981. 24 July.

Harvey, J. 1981. 'The Process of Restructuring in the Foundry Industry'. MA thesis. University of Warwick.

Lazard Frères. 1982. Press release, 23 March.

1981.

Management Today , December 1981.

National Economic Development Office (NEDO). 1977a. *Foundrymen's Views: An Attitude Study in the Ferrous Foundries Industry*. London: NEDO.

National Economic Development Office (NEDO). 1977b. *Prime Minister's Conference: The Industrial Strategy*. July. London: NEDO.

National Economic Development Office (NEDO). 1979. *Small Craft Foundries – Their Present Role and Future Prospects. A Report by the Small Craft Foundry Working Party to the Foundries EDC*. London: NEDO.

TUC. 1982. Economic Committee Paper 1/5, 13 October.

West Midlands Economic Development Unit (WMEDU). 1982. *The West Midlands Foundry Industry: A Preliminary Study*. Birmingham: West Midlands Country Council.

——. 1983a. 'Report of the Senior Adviser on Economic Development. West Midlands Foundry Sector – Summary of Respondents' Views' Birmingham: West Midlands County Council (mimeo).

——. 1983b. 'Foundry Conference – 6 May 1983. Report of Proceedings.' Birmingham: West Midlands County Council (mimeo).

Construction

In the foundry sector, government intervention aimed to foster the centralization of capital ownership and a co-ordinated approach to investment and disinvestment. In construction, by contrast, government economic policy and the competitive behaviour of firms have led to very different outcomes, characterized here as a process of 'destructuring'.

Construction was marked out as a central target for public expenditure reductions, which threw the industry into crisis well before many other sectors. Privatization and deregulation fostered destructuring, and reinforced the trend towards a new hierarchy of firms dominated by the largest construction companies which have diversified away from the competitive risks of contracting to private development and managerial services. Marginal firms have been able to remain in competition through labour intensification. Casualization of labour markets, however, has contributed to skill shortages, in spite of large-scale job loss in the industry.

One of the themes of the study is the effect of casualization and weak union organization in underwriting employer reliance on relatively backward production technologies and, in turn, undermining job security and health and safety standards. The study concludes that these are serious problems for the industry and its workforce, and for the wider economy.

3

Destructuring and Deregulation in the Construction Industry

Stephen Evans and Roy Lewis

This chapter examines 'destructuring' in construction, that is, the largely unco-ordinated, tentative and contradictory efforts of the state and capital to reorganize the industry.[1] It focuses in particular on the implications for the regulation of an increasingly casualized labour force. Casualization is a major issue not only for employers, workers and unions in the construction industry, but also for government and the wider economy. Since 1979, state policy has been characterized by 'deregulation', which has strengthened the market forces promoting destructuring. The chapter critically assesses deregulation in the context of this industry and finally addresses the question of an alternative strategy. If destructuring and deregulation are ultimately dysfunctional, what kind of institutional changes would be effective in resolving the industry's problems?

Introduction

Since 1975 the construction industry in the UK has been in a deep depression from which it had only slowly begun to recover by the mid-1980s. During this period, total output stagnated, new construction slumped, the average size of contracts diminished, product variability increased, and the geographical distribution of construction activity became increasingly concentrated in the south of England. Under the pressure of intensifying competition, the post-war trend towards industrial concentration appeared to go into reverse. Within an already fragmented industrial structure the number of small firms multiplied. Real average rates of return on capital employed (adjusting for inflation) fell from 8.4 per cent to 4.8 per cent between 1973–4 and 1982–3, well below the average for the rest of the corporate sector (Turner, 1987: 22). But appearances can be deceptive. The larger construction firms rode out the profits crisis through a number of measures. While their output fell, they enjoyed lower prices of materials and labour (Ball, 1988). They diversified away from contracting and subcontracted ever larger amounts of construction work. In the new hierarchy of contractors, the largest firms now came to dominate the market.[2] They are well placed to benefit from any revival in demand without incurring the many risks associated with large-scale direct employment. These have been passed on to medium and small firms, and especially to subcontractors and their labour.

The effects on labour of what can best be described as a process of destructuring have been equally striking. Between 1975 and 1985 total manpower fell by 0.25 million (16 per cent) and unemployment among the manual workforce reached an estimated 24 per cent in 1982. The labour force became disaggregated among many more small employment units. Labour-only subcontracting reached unprecedented levels and with it came a major shift from direct to self-employment and the casualization of the employment relationship for large numbers of workers.[3] Yet the labour market was increasingly marked by a central paradox. Alongside high unemployment, inadequate investment in training led to shortages of skilled labour in several trades, and a geographical disparity between relatively high levels of activity and demand and low skills availability. At the same time the determination of wages at local level became increasingly unco-ordinated, rendering the system of collective bargaining through national working rule agreements for the different trades ever more marginal to the actual regulation of employment conditions. Casualization thus contributed to the decline in union membership and influence. Unions became progressively less able to mobilize employers and workforce behind traditional joint regulation, or to articulate an alternative strategy for restructuring the industry. Furthermore, construction continues to have one of the worst health and safety records of any industry, a problem which the government's Chief Inspector of Factories has attributed directly to the growth of small firms, subcontracting and self-employment, as well as to the transient nature of construction work (HSE, 1987).

Even for the larger employers the rationale of destructuring is open to question. It is true that the industry's fragmented structure gives individual firms 'flexibility', and there is also some recent evidence of improvements in completion times, utilization of resources and cost control (Flanagan *et al.*, 1986). But the price for this limited progress has been high. Notwithstanding a record number of bankruptcies with the annual rate increasing from 949 to 1,975 between 1980 and 1985 (Turner, 1987: 22), the recession in construction has not shaken out many of the 'weak' firms, and the still labour-intensive production continues to be hampered by the industry's failure to provide adequate supplies of skilled labour. This may be tolerable in a recession, but quickly leads to 'overheating' when demand increases, which is now recognized as an obvious danger in the current upsurge of house, office and shop building in London's docklands and in the South East generally.[4]

In order to explain this complex picture, it may help to make some preliminary comments about the economic and social organization of the industry, and then about its relation to the wider economy and the state. The first issue to clarify is subcontracting, which is an endemic feature of the industry. Various arguments are made in support of subcontracting: it allows main contractors to bring in specialist trades as required; organizational and control costs are reduced; working capital becomes the responsibility of the subcontractor; profits can be raised by obtaining low prices through competitive tendering for subcontracts; subcontractors can be laid off without cost or serious disruption when work is delayed and, since they are smaller, closer supervision can be exercised, non-productive time and fixed labour costs minimized, and more efficient performance achieved (Hillebrandt, 1984: 120). Trade or task specialization also provide subcontractors with stability by allowing them to work on different types of product or across different subsectors of the industry. Because subcontractors tend to operate within local or regional boundaries, it is also claimed they provide greater stability of employment than the more mobile large general contractors operating within national markets (Fleming, 1977: 133–8; Winch, 1986: 112).

But this account neglects the uncertainties generated by the contracting system itself. A major uncertainty arises from the professional autonomy of the architects, their control over design, and their uneasy relationship with contractors who share responsibility for production (Ball, 1988). The resulting difficulty of accurately predicting costs in the tender price means that profit is dependent on minimizing cost overruns and maximizing claims for extra payment from the client. Moreover, each contract represents a high proportion of a contractor's total turnover so that even a small variation from the tender price can have a major effect on overall performance (Clarke, 1980: 43; Winch, 1985: 265). This has two major implications for contractors' strategies. First, avoiding risk and maintaining flexibility of financial assets are given priority over improving productive efficiency. Thus profits from contracting can be switched, inside or outside construction, to wherever they yield the best return, and working capital is kept to a minimum through subcontracting and casualizing employment (Ball, 1980: 17–30). Second, the subcontractors' relations with main contractors are subordinate, dependent and antagonistic. Subcontractors may gain a degree of independence from contractors' control if they are nominated by the architect rather than selected by the contractor, or by holding back their resources until a clear run of work is in prospect (Sugden, 1975: 74). But especially under conditions of depressed demand, main contractors can ensure control of subcontractors on site by exploiting their financial vulnerability and dependency. It is also important to differentiate between subcontracting markets, which are segmented by size of firm, product range and operating boundaries. As competition has increased, many subcontractors have had to travel further afield and have come to behave like main contractors, subcontracting more of their work to labour-only subcontractors to produce competitive tenders (Clarke, 1980: 49).

In the light of these complex developments, it is difficult to apply the notion of 'restructuring' to construction. The changes which have occurred have lacked any apparent central force or co-ordination. The crisis facing the industry began earlier and persisted longer than in many other sectors, and its origins were diverse. Of crucial importance was the state's policy of cutting back public expenditure. But also significant was the 'subjective' nature of the crisis, that is, how it was perceived within and mediated by the institutions and social relations of the industry, the firm, and the trade unions. For it was through these that sense was made of the 'objective' situation and response articulated and worked through – with the consequences for work organization and employment considered below.

Despite the industry's importance to the wider economy – in 1981 it made a net contribution of 6.4 per cent to UK gross fixed capital formation and employed 6.9 per cent of the total labour force (Hillebrandt, 1984: 4–5) – its ability to influence the level or type of demand is heavily dependent on external forces, especially the state. State-sponsored construction was central to the programme of mass housing and other infrastructures throughout the period of economic expansion from the mid-1940s to the mid-1970s, reaching a peak in 1976 of 52 per cent of all new work (Hillebrandt, 1984: 69).

Labour governments in the 1960s tried to facilitate voluntary restructuring and rationalization by private contractors through a range of incentives aimed at stabilizing demand flow, regulating competition, and guaranteeing markets and profits for firms willing to invest in industrialized construction technologies (Dunleavy, 1981: 118). By the early 1970s, however, the state's priorities had changed. Governments used the industry as a deflationary economic regulator to depress demand, reduce public expenditure, and improve the foreign trade balance (Hillebrandt, 1984: 71; Sugden, 1975: 16). From 1979 government policies towards the industry were increasingly predicated on market forces. Controls over private development, competition, and the labour market were removed, while those over expenditure on public construction were tightened (Dudley, 1983; EIU, 1981; Karn, 1985: 164; Loughlin, 1986: 136–57). By 1985 new public sector output had fallen below 32 per cent of all new work. The

Local Government, Planning and Land Act 1980 increased the proportion of public work contracted out to private contractors. For a period of time in the early 1980s, the reintroduction of fixed-cost tendering denied firms the opportunity to pass on additional wage costs. Labour standards were further attacked through measures of deregulation and new legislation hampering unions from organizing and enforcing the industry's collective agreements. As it slowly released the brakes in the mid-1980s, the government shifted greater responsibility for funding construction on to the private sector, whether through council house sales, joint private–public inner-city redevelopment, or mega-projects like the Channel Tunnel.

Given the industry's fragmented and heterogeneous character, the state's abstention from any centrally co-ordinated response to the crisis reinforced the pressure on individual firms to pursue their own short-term goals. Lower tenders and increased product variability brought further instability in the labour process, and pressures to lay off risk to subcontractors and ultimately to the labour force through the casualization of employment, culminating in the spectacular growth of labour-only subcontracting and self-employment.

Destructuring

Competition and Fragmentation
In 1985, total construction output barely matched that of 1976. New work declined by 15 per cent over the period. The sharp drop in new orders by some 40 per cent in 1975–6 had an immediate effect on contractors, the situation being relieved only by the 43 per cent increase in repair and maintenance work. From as early as 1969, the share of housing in total output fell, public sector housing being especially badly hit. This accounts for much of the decline in public sector output from its peak in 1976 of 52 per cent of all new work to below 32 per cent in 1985. The share of private industrial and commercial work meanwhile increased after 1981 from 24 to 46 per cent.

The reduction in public sector work and in private industrial work crucially affected the number of large-scale projects. Hillebrandt (1984: 81–3) estimated a fall of some 42 per cent, from 1300 to 550, in the annual number of new orders of more than £2 million between 1971 and 1981. Only private commercial work sustained its numbers of large contracts (Hillebrandt, 1984: 81–3). But here, too, was evidence of a process of fragmentation of workload, spatially and into contracts of smaller value. Orders for offices were mainly made up of small–medium sized schemes. Some 75 per cent of these were for Southeast England. The consumer boom made shop-building the fastest growing subsector of the industry, and more geographically spread as retailers decentralized locations away from town centres. In the sluggish industrial building, the newer, expanding manufacturing sectors have smaller space requirements (Turner, 1987: 20–1). Fragmentation has also been evident in the trend towards new forms of contract organization such as 'management contracting', under which the traditional single main contract is parcelled into several main or prime contracts according to the various phases and elements of the work, with each phase in turn being further divided among main and subcontractors. Alongside these developments competition for contracts and contractual conditions tightened up, especially in the public sector after 1979. Public sector tendering lists have been lengthened to include more firms, and for a period in the early 1980s central government reintroduced fixed price tenders, denying reimbursement for wage increases agreed under national bargaining and previously allowed under fluctuating cost clauses.

This recent trend contrasts with the decline in pure forms of 'open' tendering during the two decades from the late 1950s to the late 1970s. Successive government reports favoured 'selective' tendering, serial contracts, and closer working relations with fewer firms (Ministry of Works, 1944; Ministry of Public Building and Works, 1964). Selective tendering stabilized markets by dividing them into discrete segments with lists kept for different project sizes and firms' movement up to a higher rank regulated so that large firms did not compete directly with medium and small firms (Grant, 1983: 11). By 1975, only 17 per cent of building and civil engineering contracts worth over £50,000 were let by open competition (NEDO, 1975: 100–2). In this period, the construction industry lobby secured concessions from government in return for co-operation in the public housing drive. These 'included changes in tendering methods, larger contracts, continuity of work, extended control of building design, the adoption of proprietary systems, or changes in the building form used' (Dunleavy, 1981: 178).

Under the Conservative governments since 1979, policy has, in the words of Karn (1985: 164), 'been characterised by a narrower definition of the housing responsibility of local and central government; enforced privatisation of the public sector stock; draconian cuts in housing expenditure; and in particular, a style of control of expenditure which has moved from general spending targets to controls set for each individual local authority.' The reduced public sector new work, diminished size of contracts, increased proportion of contracted-out work, and the renewed pressures for open tendering have combined to intensify market competition. Large contractors have sought work of smaller value in markets previously the domain of medium and small firms. The number of small firms has increased in response to the growth of repair and maintenance work, and a squeeze has been put on medium-sized firms, which are less financially robust than the large and less flexible than the small.

Clients have also pressed their requirements more forcefully. The harsher economic climate has led them to seek more detailed involvement in the construction process, and this has stimulated the use of new contractual frameworks, 'project management' and 'design and build'. The widening range of building types and technical advances in construction methods meant that inception, briefing and design times of projects have lengthened in relation to overall construction times. Part of the attraction of these newer contracting arrangements is that clients are promised a more unified construction process, faster build times, and a more rapid return on investment, all of which help to overcome the traditional separation between design and production managements (NEDO, 1983). For management contractors the attraction is reduced risk. These considerations became more significant under the inlfationary conditions of the past decade or so because construction costs could rise appreciably in the time between inception and completion. These relatively new forms of contract have encouraged quite distinctive submarkets, dominated by the larger contractors. In 1984, management contracting is thought to have accounted for £890 million, or 10 per cent, of new non-housing work. The top 10 firms secured over half of this, and the top 16 almost 90 per cent (CCMI, 1985: 25). Between 20 and 30 per cent of private industrial and commercial work is undertaken on a design and construct basis (Hillebrandt, 1984: 23), and the proportion of public sector work undertaken on this basis is increasing.

These developments have accentuated the already fragmented character of the industry's structure. In comparison with many sectors of British manufacturing, the organization of construction has remained much less concentrated. Nevertheless, the large, mainly general building and civil engineering contractors increased in number and raised their share of workload from 15 per cent in 1935, to 31 per cent in 1954, and 39 per cent in 1968 (Carter, 1958: 47–75; Dunleavy, 1981: 15). But this trend has been reversed since the onset of the depression in the seventies. The number of large contractors with more than 600 employees halved between 1975 and 1985, while the number with seven or fewer employees more than doubled, and single person firms increased by 150 per cent from 28,131 to 72,896 (DoE, 1986: 19). This suggests a major shift in workload from large to small firms. Between 1975 and 1985, the share of work done by large (600) contractors fell from 33 to 20 per cent, and small firms raised their share from 15 to 26 per cent. One reason for this was the growth in repair and maintenance work which favours small firms. But this is only part of the explanation.

Strategies of Larger Contractors
The large contractors responded to depressed domestic markets after 1973 in a variety of ways. Apart from demanding the opening up of public authorities' own Direct Labour Organization (DLO) work to competition from the private sector (see below), the contractors' strategies have in-

cluded diversification both in and outside construction, merger and acquisition, expansion into different construction markets at home and abroad, and ever increasing reliance on subcontracting.

Diversification, which began in the 1950s, involved integration forward into property development and backward into materials and plant hire, as well as into open cast mining and off-shore drilling around the world (Channon, 1978: 155–6). Both of these trends have accelerated in recent years, and contractors have become more involved in raising funding for projects. Mergers and acquisitions have been fewer than in other industries, and have usually been confined to companies engaged primarily in construction and related activities. In private housebuilding, for example, mergers have been a means of acquiring profitable land banks (Ball, 1983: 64–79).

From the late 1960s, the international contracting activities of the leading firms grew rapidly. The value of orders peaked in 1977–8 at 11 per cent of total domestic and overseas work, falling back to 7 per cent in 1981–2. The top five contractors won nearly 70 per cent of these orders, accounting for an estimated average of between a quarter and a third of their overall workloads. However, these markets became increasingly competitive with the entry of foreign firms (Hillebrandt, 1984: 93–7) and in some mega-projects the risk of spectacular losses materialized (McKinlay, 1986: 18).

Diversification in its various forms has contributed to the development among the larger contractors of a more proactive management style. This has involved setting strict criteria for contract risk evaluation and imposing tight controls over project cost and timing. Some insights into this process are provided by a recent study of one of the major contractors (McKinlay, 1986), which expanded rapidly during the 1960s boom when it tended to regard profit as a welcome but largely unplanned by-product of increased turnover. The development of its strategic planning function was gradual, but by the end of the 1970s corporate management had secured the standardization of financial information from diverse operating divisions. Critical decisions over contract tendering and risk evaluation remained devolved, but were now circumscribed by centrally established criteria and procedures. Management's prime concern was to find and use the most cost- and time-effective construction methods, consistent with a complex set of legal obligations, to ensure maximum profitability and risk diffusion. This was achieved by 'fast track' tendering, which involved assuming full responsibility for all aspects within the terms of the contract while ensuring the firm was brought into the construction process from the earliest stage. 'Fast tracking' thus necessitated the provision of information early in the pre-tender and tender stages and a more proactive role in marketing and work-programming, with clearly defined areas of autonomy for site management. A major consequence of this whole process was to intensify the pressures and incentives to subcontract.

A significant feature of the rise of subcontracting is the increase in the share of work done by the specialist trades, which between 1975 and 1985 rose from 33 to 45 per cent. The specialist firms tend to be smaller and to work as subcontractors to main trades contractors. Until the 1970s, the main trades were predominantly undertaken by general contractors. These too are now increasingly subcontracted, in some cases to supply and fix, but more commonly to labour-only subcontractors. (Under supply and fix, the subcontractor supplies materials, plant and labour, whereas under labour-only subcontracting the main contractor provides materials and plant and the subcontractor supplies the labour and necessary hand tools.)

The greatest use and fastest recent rate of increase in subcontracting is among large general contractors (600 employees), whose subcontracted gross output rose between 1979 and 1981 from 37 to 44 per cent (Hillebrandt, 1984: 121). More recent industry-level data are unavailable, but other evidence suggests the rate has accelerated over the last few years. Within our own investigation of 12 national and regional contractors, the weekly labour returns for all sites in a single operating region of one large contractor from 1982 to 1985 showed that the proportion of the firm's own direct employees fell from 35 to 20 per cent, its supply and fix subcontracted workforce fell from 46 to 35 per cent, whereas its labour-only subcontracted workforce increased from 19 to 45 per cent. Shifts at least as great were recorded among the other firms. The trend is confirmed by other researchers (Bresnen *et al.*, 1985: 112; Clarke, 1980: 43–7).

Changes in the Labour Market
These changes in the economic organization of the industry have had profound implications for many aspects of the labour market, including employment levels, labour mobility and the growth of self-employment.

There have been massive job losses. Between 1975 and 1985, manpower declined by 255,000 to just under 1.5 million. The government stopped recording unemployment within the industry in 1982, when it stood at 360,000 or 24 per cent of the total, having risen from 11 per cent in 1979. Union estimates put the level at 450,000 or 30 per cent in 1985 (UCATT, 1986). These figures mask regional variations, stimulated by the 'increasing imbalance in the levels of construction activity between the north and south' (NEDO, 1986: 2). Between 1980 and 1985, directly employed labour declined by an average of 12 per cent in Greater London and the South and 31 per cent elsewhere, except in the North where it fell by 52 per cent (DoE, 1986). At the same time, the labour force became more disaggregated with the increase of subcontracting among large contractors. Their share of employment fell by 11 per cent to 22 per cent between 1975 and 1985, and that of specialist trades firms rose by 10 per cent to 45 per cent.

Labour mobility, voluntary and involuntary, has also increased. Recent survey evidence found that 42 per cent of the manual workforce had changed jobs at least twice during the previous five years, and 41 per cent had experienced unemployment of at least six months over the same period. Recruitment from outside industries and exit and return by construction workers were similarly high, a process which depresses conditions for the less skilled especially (Marsh *et al.*, 1980: 5–6).

Administrative, professional, technical and clerical (APTC) staffs in private organizations have, however, remained relatively stable through the recession, declining by just 9 per cent between 1975 and 1985 and by slightly more in the public sector, compared with a fall of 31 per cent of employed manual workers. APTC staff account for 29 per cent of all construction employees, and over one third of large contractors' labour forces. They have become a major 'fixed cost', contributing to the pressures to casualize on-site labour.

Self-employment is the classic symptom of the casualization of labour in the construction industry. The spread of labour-only subcontracting in various forms – from agencies supplying labour, to gangs and individuals – inexorably led to the proliferation of self-employed operatives. Many of these workers are arguably only nominally self-employed as they work exclusively for contractors, providing only their labour and hand tools, and are therefore distinguishable from genuinely self-employed tradesmen in business on their own account. Legally, however, labour-only workers occupy 'a twilight zone between employment and independent status' (Chesterman, 1982: 38). They frequently lack a contract of employment with main or subcontractors.[5] While directly employed manual operatives often fail to qualify for statutory rights because of insufficient continuity of service, the absence of an employment contract means that contractors automatically avoid statutory rights to notice periods, redundancy pay and protection from unfair dismissal. Furthermore, they have no need to pay either 'stamps' under the industry's holiday pay scheme or National Insurance contributions, and they are subject to less onerous obligations than employers in respect of accident compensation and insurance.

The most tangible benefit for the self-employed worker – apart from allegedly higher pay – is a tax advantage. Under the '714' certification scheme introduced by the Finance Act 1971 as amended, an operative may obtain a 714 certificate if he can satisfy the Inland Revenue that he is genuinely self-employed. The criteria include a satisfactory record of tax and National Insurance payments, the normal assets of a genuine business (premises, equipment, etc.), and a public liability insurance policy, though recently the insurance requirement has been diluted as part of the deregulation policy (see below). The possession of a 714 certificate means that PAYE income tax is not deducted at source and instead the workers' profits (fees less the quite generous scale of Schedule D expenses) are subject to an annual tax assessment. If an operative lacks a 714 certificate and is not directly employed, then the contractor must deduct a percentage of his earnings at source – currently 27 per cent – and send it to the Inland Revenue.

The most reliable indicator of the scale of self-employment is the number of construction workers in possession of 714 certificates. Between 1978 and 1985 the number of certificates issued rose from 283,692 to 452,415. Seventy per cent of these were issued to individuals. Data for the period 1964 to 1981 suggest a high point for labour-only subcontracting on a self-employed basis of 27 per cent in 1978–9, declining to 22 per cent by 1981. This is almost certainly an underestimate, since it excludes workers believed to be serving the period of qualification for the 714 certificate and who should have arranged for the contractor for whom they were working to deduct 27 per cent of their earnings. These would add a further 150,000 to the registered 714 certificate holders, making a total for 1982 of around 600,000. This compares with an estimate of under 200,000 in the 1960s (Phelps Brown, 1968: 114–17).

Labour Regulation

The recent trend towards the casualization of construction labour contrasts with and yet evolved from the rather different employers' labour and industrial relations policies which held sway during the post-war expansion, especially in the boom years of the 1960s. At that time, contractors tried to avoid work programming problems and meet fluctuations in demand for different trades through employment policies predicated on securing a regional workload sufficient to allow for the retention of a nucleus of direct labour (Pinschoff, 1970). Indeed, the provision of stable employment for local labour was often a condition of winning contracts from public sector clients. But high levels of demand also ensured that firms generally had to take whatever labour was available, and this led to an emphasis on control rather than planning as the central problem for site management. On site, where one *force majeure* followed another, the main issues of control were variability of worker performance, work tempo and the phasing of new work stages (Sugden, 1975: 76). Such problems combined with product instability to help perpetuate labour-intensive production methods, detailed direction of which remained largely under the control of the workforce. This in turn accentuated the difficulties of raising labour productivity, and highlighted the advantages of direct employment in assisting that objective (Carter, 1958).

Growth of Labour-Only Subcontracting
Labour-only subcontracting has always been a feature of the industry, especially in speculative housebuilding. But in recent years it has expanded rapidly in other sectors as competition intensified, the size of contracts diminished, and the opportunities to cheapen and casualize labour in-

creased. For the larger contractors labour-only subcontracting has great economic advantages. Several firms in our sample of contractors calculated that tenders based on subcontractors' labour prices provide savings over direct labour costs of around 20–30 per cent. It was only the mediation of social relations within firms which delayed the pace of change. The practice of carrying a nucleus of favoured regulars and its supportive managerial culture, a blend of paternalism and craftism, meant that many managers 'knew no other way' than direct employment. The cost of redundancy payments, the practice of transfers while order books were full, the risk of disputes, and lack of good quality labour among subcontractors all served to postpone a sudden shift to wholesale labour-only subcontracting.

However, as subcontractors came increasingly to perform the labour supply function for large contractors (Bresnen *et al.*, 1985: 117), the need for labour mobility proved incompatible with stable employment and there was a resurgence from the late seventies of labour-only subcontracting. It spread deeper and more pervasively among main trades (its traditional base) and among the specialist trades such as electrical contracting where it had previously been marginal. A survey by the Electrical Contractors' Association of 41 per cent of its members in 1986 revealed that 36 per cent of man–days' labour was carried out by self-employed labour-only subcontractors of various kinds (Cathcart, 1986). It has also become more widespread of late in public sector projects (Clarke, 1980: 46).

The growth of labour-only confounds the prediction that it would decline as demand fell off and the opportunity for top craftsmen to 'bonus chase' from site to site diminished. According to Phelps Brown (1968), supply-side pressure was the major stimulant, together with the well-known Schedule D tax advantages for the self-employed worker. After the introduction in 1966 of Selective Employment Tax, increased fixed labour costs encouraged employers and workers to collude in order to avoid state imposts (Undy *et al.*, 1981: 282). Another factor was the effect of industrialized building and labour-only in changing the requisite skills for site management – closing off opportunities for promotion from craftsman to foreman and site agent, and depressing earnings differentials (Morton, 1979: 15). Some commentators prefer to stress demand-side pressures, especially the role of large contractors' employment policies. These firms are mobile within national markets, whereas labour is comparatively immobile. Working only irregularly in any one area, firms nevertheless seek labour with a continuous work history, ready trained for specific tasks, and recruited on recommendation via established 'grapevines'. In this way, those workers initially attracted to labour-only subcontracting for its high rates become 'trapped', with little chance to widen their skills and thus be in an advantageous position to secure direct employment (Winch, 1986: 110). None of these explanations, however, places sufficient emphasis on developments in production and management organization or on the decline of joint regulation.

The Management of Casual Labour

Managements have limited the effects of product instability and labour process variability through more effective systemization in the estimating and work-programming stages of projects (McGhie, 1982: 3). According to management in one of the top five contractors (part of our sample of large contractors), every job is now tendered for quoting subcontractors' prices. Estimaters check these prices against a large database of standard times or 'elements' calculated for each task. Work programmers use these standards to define schedules against which site managers determine their labour requirements. This process has been facilitated by the reduction in architects' autonomy and the increasing use of standardized, factory-produced components.

Another aspect of the changing function of management is indicated by the modifications to the industry's standard forms of commercial contracts (see, generally, Uff, 1985: chs 10–12), which have been introduced at the initiative of clients and larger contractors under the new conditions of 'fast tracking'. They are drafted to ensure that subcontractors commit adequate resources and assume financial responsibility, though they do not appear to prevent a considerable volume of litigation and commercial arbitration. In fact, the practical limits of relying on the formal contractual provisions are indicated by frequent resort to informal economic pressures on subcontractors (see Lewis, 1982) and the maintenance of approved lists of subcontractors – a practice designed to cultivate co-operation, flexibility and access to supplies of labour. Generally, as the role of large contractors changes to one of managing contracts as opposed to employing labour, subcontractors will have to improve their expertise in managing labour. Thus, it is common now for contractors to insist on subcontractors allocating particular foremen to projects as a condition of being awarded the work.

Labour is deployed in different ways to suit specific project conditions. In some cases, it is interchangeable and work gangs are allowed to arrange themselves for the various tasks, in others workers are allocated to specific tasks by foremen or higher management. This flexible approach builds on the experience gained in the 1960s of increasingly narrow subdivision and specialization of tasks within trades, and appropriate matching of labour of varying abilities (Pinschoff, 1970). One objective is to avoid breaking up the 'social solidarity' of regular work gangs, which in the past encouraged an unnecessarily high rate of voluntary labour turnover (Allen, 1952). Subcontractors tend in practice to retain a nucleus of regulars supplemented as necessary by casual hirings. But as labour-only subcontracting spreads, the form of employment – that is, contract of employment for direct employees and contract for services for self-employed – provides few pointers to the real substance of the employment relationship. According to a senior industrial relations manager in one of the top five contractors (part of our sample of firms), the typical subcontractor's labour force may now comprise direct, 714, and '27 per cent' labour in roughly equal proportions. The rising number of self-employed implies deteriorating security for many former direct workers. For others, it represents merely a formal change of status rather than a substantial change in their relations with the employer (Austrin, 1980: 300). Yet the demise of skills training, combined with 'overheating' in certain trades and localities, produces skill shortages from which some workers clearly benefit. On the whole, however, the labour force comprises an ever growing number of workers trapped in casualized, narrow skill-based jobs.

The System of Joint Regulation
Labour in construction is formally regulated through national collective bargaining based on detailed Working Rules for the various trades and sub-sectors. But this framework has become increasingly peripheral, especially among the subcontracted labour force. Conditions of employment are now often influenced by local labour-market conditions and imposed by employers or bargained with individual workers.

The major agreements are negotiated at national or industry level in joint industrial councils or boards made up of the relevant employers' associations and trade unions. The 'building' trades are peculiar in that they have two competing sets of agreements. The National Joint Council for the Building Industry (NJCBI) with its National Working Rule Agreement (NWR) was the first established, and carries considerable weight, especially among the general contractors. The NJCBI parties are the Building Employers' Confederation (BEC) and four unions: Furniture, Timber and Allied Trades Union (FTAT), General Municipal and Boilermakers (GMB), Transport and General Workers' Union (TGWU), and Union of Construction and Allied Trades and Technicians (UCATT). The other agreement was established in 1980 through the formation of the Building and Allied Trades Joint Industrial Council (BATJIC), which is negotiated between the TGWU and the Federation of Master Builders (FMB), which represents mainly small contractors.

In Civil Engineering, the Civil Engineering Construction Conciliation Board (CECCB) Working Rule Agreement is negotiated between the Federation of Civil Engineering Contractors (FCEC) and three unions, GMB, TGWU and UCATT. 'Large site' heavy engineering construction is also covered, since 1981, by the National Joint Council for the Engineering Construction Industry (NJCECI), which has a National Agreement for the Engineering Construction Industry (NACEI) – bringing together several different trade agreements – negotiated by the National Engineering Construction Employers' Association, Oil and Chemical Plant Constructors' Association (OCPCA) and Thermal Insulation Contractors' Association (TICA) with six unions: Amalgamated Engineering Union (AEU), AEU (Construction Section), Electrical, Electronic, Telecommunication and Plumbing Union (EETPU), Technical and Supervisory Staffs (TASS), TGWU and GMB. The largest specialist trade is electrical contracting, which is covered by the Electrical Contracting Joint Industrial Board (ECJIB), representing the Electrical Contractors' Association (ECA) and the EETPU. The same union negotiates exclusively with the National Association of Plumbing, Heating and Mechanical Services (NAPHMS), under the Joint Industrial Board for the plumbing industry. The Heating and Ventilating Contractors' Association (HVCA) negotiates the National Agreement of the Heating and Ventilating, Air Conditioning, Piping and Domestic Engineering Industry exclusively with TASS.

The different trade agreements vary in detail, but share certain general characteristics. They set national pay rates, which in some cases are regarded as maxima and in others (like building) as minima. They also specify payments for special conditions and tasks, hours and 'fringes' such as holiday and sickness credits, provide procedures for introducing bonus schemes, and establish rules on labour mobility, training and grading qualifications. It is especially noteworthy that they impose duties on contractors to ensure that subcontractors comply with agreements and prohibit self-employment (Gallagher, 1984; Hilton, 1968).

Comprehensive, trade or industry-wide standards and centralized barganing were developed from the turn of the century by the small specialist contractors in the different trades. The objective was to provide stable and predictable labour costs for tendering purposes, take wages out of competition in local markets, secure union participation in the supply of standard labour skills and in disciplining labour militancy on site, and, crucially, to regulate competition and undercutting. Since then, adherence to the collective agreements among employers and the cohesion and disciplinary authority of their associations has varied from sector to sector, reflecting the tensions among different firms according to size, market position and labour policies.

The rules were most tightly enforced where the trade was relatively homogeneous and union density was high – for example, in electrical contracting, the firms enjoyed comparative protection from competition through nomination by architects in an expanding market. Building, by contrast, had a more complex 'ecology' of subsectors and trades (Lange and Mills, 1979: 1), areas of which, such as private housebuilding and repair and maintenance, were traditionally hardly touched by the NJCBI agreement, since many of the developers and contractors did not belong to the National Federation of Building Trades Employers (NFBTE), the forerunner of the BEC. The advent of the BATJIC agreement has remedied this to some extent, but its coverage is still restricted. Furthermore, some of the new trades associated with new materials and technologies are not covered by existing agreements, and the expanding numbers of subcontractors have generally been inadequately represented within the policy-making procedures of the employers' association. But most importantly, the growth in influence of the large general contractors made national bargaining less relevant. With union strength patchy and limited to the main conurbations, these firms favoured decentralized collective bargaining, which enabled them to compete for adequate supplies of labour in the face of local labour shortages and, later, incomes policies.

Decline of Collective Bargaining
Gradually the formal bargaining institutions became marginalized. The two-tier (craft and labourer) pay structures in building, for instance, failed to adapt to technological change and the increaed demand for semiskilled labour. In addition, the practical difficulties inherent in the application of nationally agreed criteria for incentive bonus schemes at site level encouraged employers to continue to make use of traditional 'spot' bonus or plus rates and guaranteed overtime in order to raise productivity and attract labour (NBPI, 1968: para 67). The effect was to turn differentials 'upside down' (Morton, 1979: 91; Routh, 1980: 127).

Such modes of payment are an important part of the explanation of the growth of labour-only subcontracting and self-employment. Motivated by a simple incentive based on spot bonus and traditional piece rates, self-employed gangs were able to work faster and earn more than direct employees (Gagg, 1969: 132). But the price was fixed only for the particular task or contract, which as far as employers were concerned avoided problems of consolidation. Nowadays, many subcontractors pay their direct employees a straightforward shift payment based on the local 'going rate', which includes travel and overtime payments (calculated below the collectively bargained rates), and excludes wet-time, holiday, sickness, and other payments guaranteed under the working rules. Payments to labour-only subcontractors are calculated on the same all-in basis, sometimes by the piece, sometimes on an hourly or shift rate, with an extra percentage over direct workers' rates to meet their overheads.

For a long time the employers' organizations warned that an unregulated market could depress tender prices, 'jack up' pay rates, reduce union authority over site militancy and expose contractors to being picked off on prestige sites. But individual firms increasingly breached the working rules by engaging, or allowing their subcontractors to engage self-employed, labour-only subcontractors. As union influence continued to decline, the 'lump' (as it came to be called) achieved legitimacy through the 714 scheme and many contract (commercial) managers found it increasingly difficult to reconcile the reality of self-employment and the 714 scheme with the prohibition of self-employment in the working rules. The effect was to downgrade still further the importance of industrial relations considerations in the process of contract allocation and enforcement.

As far as the construction unions are concerned, the growth of self-employment exacerbated a number of long-standing problems. Membership and bargaining power were in decline throughout the post-war years. Union wage policies failed to adapt to the changing situation, despite some ephemeral gains from wage militancy at site level. The unions did reorganize themselves to some extent, but could not overcome rivalries between national and local levels and between different unions. Moreover, they were unable, if not actually unwilling, to mobilize opposition to employers' policies outside the industry in the broader labour movement. Although the policies of Labour and Conservative governments were not, of course, identical, the state remained unsympathetic to union demands in the 1960s that labour-only subcontracting be prohibited by statutory means, and union efforts in the 1970s and 1980s concentrated on the narrow aim of trying to bolster up diminishing employer support for joint regulation.

Labour turnover and mobility always made union recruitment and solidarity difficult to sustain in construction. The closed shop was not formally recognized in working rule agreements, although in practice 100 per cent union membership was more common in major centres and on large sites (McCarthy, 1964: 65), but otherwise employers successfully resisted it (Burgess, 1975; Price, 1980). Even during the 1960s boom, and against the trend elsewhere, union density declined. From 1948 to 1974, union density overall fell from 45 to 27 per cent, and from 1955 to 1971 the craft unions, which came to form UCATT, lost 31 per cent of membership (Undy *et al.*, 1981: 150). Most recently, the revival of construction activity in London and the Southeast has led to a modest reversal of this trend.

Given their dependence on employer support, national union leaders tended to favour national bargaining. It covered the weaker as well as the stronger areas and ensured unions a role in industrial regulation. By establishing relatively exclusive bargaining jurisdiction for individual unions within their separate trades, these arrangements gave protection from predatory membership raids, but they also made the unions slow to adapt or to organize the new semiskilled jobs (McCormick, 1980: 183–6). Official wage policies, especially in building, clung to national negotiations, and this meant that national rates failed to keep pace with site earnings, maintain craft differentials, or match labour-only subcontracting rates (Undy *et al.*, 1981: 176).

Unions in Conflict
These policies eventually brought the national and local union organizations into sharp conflict as the latter tried to fill the gap at site level. In the 1960s, shop stewards pressed for site rather than gang bonus, treating bonus as 'guaranteed' and the basic rate as attendance money. In some cases they imposed restrictions on hiring and firing, output and working in wet weather. As economic conditions began to deteriorate, the focus of bargaining shifted to job security, transfers and redundancy, with stewards having some success in using inter-site organization to restrict firms' efforts to start up jobs with new (subcontract) labour. In fact, there was relatively little site negotiation and stewards were thin on the ground in the industry as a whole even in the 1960s (Undy *et al.*, 1981: 176). But they were

common on large sites, and there was 'hardly a firm of substantial size that did not have to contend with the effects of militancy on one site or another' (Morton, 1979: 34). From 1947 to 1973 construction was the most strike-prone industry after coal mining, measured by the number of strikes. Between 1960 and 1973 (and excluding 1963–4 and 1972 when there were official national disputes), all but 9 per cent of strikes measured by working days lost were unofficial, and in the four years up to 1971 the annual average of working days lost due to strikes per 1000 employees rose from 125 to 200 (Durcan *et al.*, 1983: 176, 193). Unofficial militancy exposed the national union leaders' lack of authority, undermining their demands in particular for the imposition of sanctions on contractors who broke working rules by using 'lump' labour. It also persuaded many employers (outside electrical contracting) of the industrial relations advantages of subcontracting. In contrast, the electrical contracting JIB was devised precisely to allow the strongly anti-communist electricians' leadership to control local militants.

In the 1980s, depressed demand and the relentless rise of self-employment have further depleted the ranks of shop stewards. Our research indicates that in 1985 in UCATT's London region, where demand remained relatively buoyant and where there were still large sites, there were just 13 convenor stewards among private contractors, and only one worked for a subcontractor. In the London and Eastern regions there were only 43 stewards employed by subcontractors. Employers have refused to modify the working rules to allow for site-wide steward organization on management contract projects, and stewards have gradually become more amenable to the control of both management and full-time union officers.

Inter– as well as intra-union conflict has become a feature of construction industrial relations. Historically, the different conditions facing unions in their respective trades made any prospect of one union for construction remote (Druker, 1980). UCATT is an amalgamation of formerly separate craft unions, and claims to be *the* union for building workers. A similar claim is made by the TGWU which organized labourers and semiskilled workers until the craft-based National Union of Plasterers and the Scottish Slaters and Tilers transferred engagements to it. From the early 1970s, the continuing loss of members and revenue brought the two unions into sharp conflict as they opened up their ranks and attempted to recruit all types of building labour, irrespective of status.

These conflicts vitiated union efforts to resist labour-only subcontracting and other manifestations of casual labour. The National Federation of Building Trades Operatives, the union body which formerly co-ordinated industry-level bargaining until it was disbanded in 1969, always deprecated unilateral action. Individually, however, national unions pursued joint or unilateral regulation depending on market conditions, organizational resources and pressures from local level. Local activists and shop stewards stressed site militancy, strikes, blacking and picketing, allied with enforcement of the closed shop and influence through sympathetic local authorities.

Working Rule 26

These diverse pressures ultimately focused on the national working rules prohibiting labour-only subcontracting. These had evolved from 1950 when a 'settlement' safeguarded *bona fide* subcontractors. The problem was then seen largely in terms of subcontractors who defaulted, leaving workers without wages or other benefits. In 1951, criteria for such subcontractors were specified. These included financial substance and business status, use of NFBTE standard forms of subcontract, engagement of apprentices, and affiliation to the NJCBI. The remedy for breach of rules – for example, defaulting on wages or holiday pay stamps – was to refer the subcontractor to the NJCBI Joint Conciliation Panel. In 1964, Working Rule 8 placed responsibility for subcontractors' conduct on the main contractor. The unions disagreed among themselves over this provision. The Amalgamated Union of Building Trades Workers, the bricklayers, objected to it on grounds that it legitimized the 'lump', a view with which many employers concurred. But it did unambiguously prohibit self-employment. It was replaced in 1971 by the Declaration of Intent that all operatives should be in the direct employment of the contractor or subcontractor and employed under the conditions laid down by the NJCBI. However, unions pressed further and after the national strike in 1972, the employers felt compelled to 'give a bit' with the introduction in 1973 of Working Rule 10, later to become National Working Rule 26. This reiterated the principle of direct employment embodied in the Declaration of Intent, and provided for deduction of union dues and notice of new contracts, and union access to subcontractors. NWR 26 applied only to firms who voluntarily registered with the NJCBI. Most were initially reluctant to register, but did so once they came under pressure from public sector clients who were persuaded by the unions to insist on registration as a condition of entry on to approved tender lists.

The remedies available against self-employment under the working rules of the various trades (not just building) include removal of the subcontractor from the job, replacement of the self-employed by direct employees, and expulsion of contractors from the register. The removal of the subcontractor used to be the preferred union approach, although there was no assurance that those removed would be replaced by a *bona fide* firm. Instead, unions came to favour the replacement of self-employed workers by direct employees and 'educating' the main contractors to assume their responsibilities. Unions argue that these extend to labour engaged by any subcontractor wherever they are working, but employers insist they apply only to the site in question. The cost and limited availability of 'good' direct employees has made contractors increasingly reluctant to accede to union arguments. Finally, there is apparently not a single case of expulsion from the NJCBI register (and just one instance under the Electrical Contracting JIB), unions preferring the threat rather than the reality of expulsion for fear of inciting open employer opposition.

In fact, the conciliation machinery is only rarely invoked on this issue, though our research indicates that there has been an increase of late. Industrial action or its threat has similarly been used occasionally, and remains a potent if diminishing deterrent in well-organized trades and localities, though it may be subject to legal restraint.[6] Campaigns against the 'lump' continue intermittently, but are moderated by unions' increasing dependence on the goodwill of the large contractors. The unions' ability to influence contractors' selection of subcontractors and to enforce the working rules has diminished as industrial relations considerations have been subordinated to price and performance criteria. Public sector clients, too, are more reluctant to put pressure on contractors to ensure that NWR 26 is observed. At the same time, the role of commercial contracts in the enforcement of collectively agreed terms and conditions has been jeopardized by legislation, which we discuss below.

Decasualization or Deregulation?

Discussion of destructuring and its effects in eroding joint regulation and intensifying the trend towards the casualization of labour raises the question of whether there can be an alternative strategy for the industry. This, in turn, requires an examination of the approaches of unions and workers, employers and the state.

Union Policies

Within the labour movement, three main lines of prescription have been advanced: taking the contractors into public ownership and expanding local authority Direct Labour Organizations (DLOs); using the state's economic and contractual powers as client to compel contractors to hire labour direct; and introducing registration of employers and workers, either on a voluntary basis jointly regulated through the industry's collective bargaining machinery, or by statute.

Construction unions have long recognized the state's potential role in restructuring the industry. But they have been ambivalent about demanding direct state control, among other reasons because of the daily reality of trying to sustain employer support for joint regulation. Union demands always fell short of wholesale nationalization. They pressed the Labour Party to take the larger contractors into public ownership, moderating this to proposals to the 1977 Labour Party conference and the 1981 Trades Union Congress to set up a National Building Corporation (NBC) to act as a procurement agency for public sector contracts. The NBC would have operated on a regional structure to ensure that the workload was evenly spread and resources used to best advantage. Its intended role, however, was somewhat ambiguous. On the one hand, it was to plan and manage public sector work to help lower unit costs and attract the larger and more efficient firms to work exclusively for the NBC, thereby maintaining continuity of work and employment. On the other hand, it was to compete for private sector work, by developing, co-ordinating and standardizing DLO construction techniques across regions and between the NBC and local Enterprise Boards. But the unions in effect dropped these proposals and have tended to press most strongly for contract compliance policies linked to measures – either voluntary or statutory – to decasualize the labour market through the registration of employers and the certification of workers' training achievement. This strategy depended ultimately on the adoption of favourable attitudes towards organized labour by the state and by public sector clients, since the support of workers and employers for tighter regulation remained ambivalent.

Worker Orientations
Workers' attraction to proposals to decasualize by standardizing and regulating the supply and conditions of labour have varied depending on their position in the labour market. For the fastest workers, also commonly the younger ones, what is seen as the union rate may be too low, and detailed regulation may be perceived as a threat to the relatively higher earnings they can make on a self-employed basis. The least skilled may likewise be antagonistic to tighter controls over access to skill acquisition and employment. Both groups are likely to resent the stricter discipline imposed by direct employment over their freedom to search for the best jobs. However, recent survey evidence shows that many workers would prefer more stable employment and favour the introduction of skills certification. Two thirds of the industry's manual workers claim skilled status, but less than one third possess any formal qualification. 'Certification would assist skilled men to get the jobs they want and would help to keep out unskilled operatives' (Marsh *et al.*, 1980: 3).

Employers and Decasualization
Employers traditionally opposed the extension of public ownership and DLO activities, and were prepared to accept registration and certification only on a voluntary basis. Such schemes have operated under the working rules of the specialist electrical and engineering services sectors, and were introduced in tunnel mining, scaffolding and demolition within the last 10 years. But the demolition register was wound up in 1984, and labour is seldom recruited from registered lists in any of the trades. As a general rule, employers have resisted proposals which threatened to restrict their access to specialist, ready-skilled labour, to deprive them of flexibility to motivate labour by means of 'spot' payments and casual hiring and firing, or to encourage union controls, the closed shop and site militancy.

State Policies Before 1979

Post-war state policies have consistently endorsed self-regulation by the industry. All governments seem to have been dissuaded from directly reorganizing the industry on account of its sheer scale and the logistical problems and costs of its administration. Even the DLOs, the prime illustration of direct state intervention in the 1960s and 1970s, represented only around 11 per cent of fixed construction output throughout the period (Hillebrandt, 1984: 74).

State policies towards construction labour markets likewise tried only to steer the industry's own institutional mechanisms in the direction of broader governmental goals. In practice, these policies were largely unco-ordinated, reflecting the conflicting objectives of different government departments. In 1966, for instance, the logic of shaking out surplus labour from manufacturing industries required that the construction industry bore the full brunt of the Selective Employment Tax, which – together with the Redundancy Payments Act 1965 – merely strengthened the forces pushing labour from direct to self-employment. This in turn contributed to wage inflation and losses of tax revenues from bogus self-employment. These developments as well as several major strikes prompted a series of inquiries into the industry's employment conditions and labour relations.[7] But in all these inquiries, the issue of decasualization was treated as subordinate to the needs of employers and the Treasury. Governments were consist-ently advised to regulate rather than eradicate labour-only subcontracting, and to avoid compulsory or statutory registration. Thus the Phelps Brown Report emphasized the economic advantages of labour-only subcontract-ing and recommended voluntary registration of genuine employers with adequate insurance cover and of genuinely self-employed tradesmen. It also proposed to give greater legal protection to the non-registered self-employed by extending the liabilities of both main and subcontractors.

The Labour government accepted Phelps Brown's overall endorsement of labour-only subcontracting and in 1970 introduced the Construction Industry Contracts Bill.[8] The Bill embodied a legal framework for the voluntary registration of contractors, including a system of financial im-posts on non-registered persons who might otherwise have failed to meet the obligation with regard to tax and accident liability insurance. The Bill received support from all Parties but was opposed by the Treasury and Inland Revenue because it cut across their own plans to tackle tax abuse. In the event, the Bill fell when Labour lost the 1970 election. The Treasury's proposals for a tax deduction scheme specification for construc-tion contractors were enacted in the Finance Act 1971.

The minority Labour government of 1974–6 resisted demands for the reintroduction of the Construction Industry Contracts Bill and instead established a new advisory body, the Construction Industry Manpower Board (CIMB). This was constituted by equal numbers of employer and union representatives under an independent chairman, and espoused the traditional cause of joint regulation, asserting the advantages of voluntary over statutory registration schemes in securing employers' support. It advocated criteria for employer registration which mirrored those of the 714 scheme, and proposed that possession of a 714 certificate should qualify for registration (CIMB, 1980: paras. 2.7.1. – 2.7.6). In this, it followed Phelps Brown in treating labour-only subcontractors as normal employers, with the potential for employing their own labour directly (cf. Phelps Brown, 1968: 156–7). However, the CIMB was never given any independent authority by government, and had no sanctions to enforce its registration proposals. In the end, it was abolished by the Conservative government in 1981.

State Policies in the 1980s

After 1979, the prospects for decasualization became even bleaker. Conservative policies had four main strands. First, the industry was used to depress demand, with a major reduction in public expenditure on housing and civil engineering in particular (EIU, 1981). Second, the DLOs, which had already been targetted for cuts under Labour, were curtailed by the Local Government, Planning and Land Act 1980 and regulations made thereunder: specifically, they were required to operate on a commercial footing and to compete for tenders with private contractors for an increasing volume of local authority work (Loughlin, 1986: 152–4; IPM/IDS, 1986: 18–20). Private contractors whose opposition to DLOs had sharpened during the recession (Langford, 1982: 25), benefited considerably from this enforced contracting-out (Layton, 1985: 10; Ramsdale, 1985: 7). In future, the principle of compulsory competitive tendering is to be generally extended by the Local Government Act 1988. Third, deregulatory measures aimed at promoting competition in the labour-market were introduced, and fourth, changes in employment law were enacted to undermine collective bargaining and union activity.

The measures to deregulate the construction industry began in 1980, when the Department of Environment (DoE) notified the NJCBI and the local authorities that it was withdrawing support for the NJCBI Declaration of Intent on direct employment. In a move regarded by many on all sides of the industry as having directly stimulated the resurgence of self-employment, the Finance Act 1980 abolished the requirement that 714 certificated labour-only subcontractors be insured against liability as an employer and to the public in a sum not less than £250,000. In 1981, the government wound up the CIMB, and in 1983 the long-standing public policy of using central and local government to set labour standards through their contracts with private contractors was reversed with the rescission of the Fair Wages Resolution. The repeal of schedule 11 of the Employment Protection Act 1975 was intended to have a similar effect in frustrating the wider application of recognized terms bargained at national level or, where these were absent, of general terms equivalent to the 'going rate' in the locality or trade.

At the same time, novel legal restrictions were imposed on the use of private contracting power to support union membership and recognition. The Employment Act 1982 ss.12–14 outlawed clauses in commercial contracts – including contracts for the services of workers who are legally self-employed – stipulating the use of unionized labour or recognition of a union: employers enforcing such clauses are liable to commit a tort against the party who is excluded from a tender list or who is refused or removed from a contract, and unions lose their immunities for industrial action taken to uphold union– and recognition-only practices (Davies and Freedland, 1984: 667–70; Lewis and Simpson, 1986: 72–4). While contractual stipulations for unionized labour or union recognition are clearly outlawed, the 1982 Act may also invalidate the terms of a contract – between a main and subcontractor or client (such as a local authority) and contractor – requiring the application of NWR procedures including, in particular, NWR 26. Furthermore, the Local Government Act 1988 has prohibited local authorities from taking into account in the tendering process 'non-commercial matters', including terms and conditions of employment, whether a contractor engages self-employed labour, and his attitude towards trade disputes.

Some employer organizations raised objections to the 1982 Act. They were worried that it would lead to contractors being sued by subcontractors looking for a 'fast buck' if they were excluded for non-compliance with NWR 26, and that it would allow non-recognized, predator unions entry to established bargaining arrangements. They complained also that deregulation would encourage 'cowboy' firms to undercut tender prices, stimulate wage inflation and deter investment. Most of these objections, however, emanated from managers concerned directly with industrial relations. In fact, the very provisions of the 1982 Act which worried the industrial relations professionals were introduced largely in response to pressure from construction employers. Business leaders in the industry generally took a broad view and endorsed the government's overall programme. While they pressed for domestic construction activity to be expanded, they were content to boost their profits with overseas activities facilitated by the lifting of exchange controls. In practice, the legal restrictions have not adversely affected employers. The large firms' growing involvement in management contracting and their move away from large-scale direct employment has meant that they are less affected by deregulation and increased competition. The subcontractors' economic dependence on the larger contractors has in turn ensured that they have only very rarely threatened litigation against main contractors. At the same time, the 1982 Act benefited contractors to the extent that local authorities felt constrained because of the risk of litigation from taking action against them when the unions raised the issue of breaches of NWR 26. In future the Local Government Act 1988 is likely to deter local authorities from making any requirement on contractors to observe NWR 26.

Conclusions

The recession appears on first examination to have wrought few fundamental changes on the construction industry. It is clear that the tendency towards fragmentation has intensified through the proliferation of small firms, the increasing volume of subcontracted work, and the relentless growth of self-employment. But just beneath the surface changes of greater significance for the future character and performance of the industry have occurred. We have tried to encapsulate these development in the concept of destructuring. The reduction in volume and scale of activity with the consequent intensification of competitive pressures, the 'privatizing' of construction work hitherto the responsibility of state agencies, and the growing client involvement in project management have helped to transform the large contractors from direct employers and producers to private developers and providers of managerial and technical services. The risks and uncertainties of the system have meanwhile been shouldered by the ever-increasing number of subcontractors and by construction workers. The largest contractors are now well placed to capture a growing share of the next cycle of expansion, and their strong position has been reinforced by state policies of contracting-out and deregulation. Although labour deregulation has been a less important factor than reductions in public expenditure and tighter local authority budgetary controls, it remains a potent constraint on a possible revival of union opposition to the power of private contractors, whether through collective bargaining or via pressure for state intervention, particularly at local level.

Our analysis of destructuring has shown it to have had a multiplicity of causes. Changes in the pattern of demand have been of major importance, but they do not provide a total explanation. Variability of product and hence of the labour process, have commonly been identified as major sources of uncertainty for construction employers, who have consistently pressed the state to ensure more stable demand. The state's response in the 1960s was to experiment with restructuring the industry along 'Fordist' lines, giving contractors incentives to rationalize and invest in industrialized systems. The failure of this experiment, however, should not detract from the continuing efforts of suppliers and contractors to seek a comparable solution to problems of variable demand.

Destructuring has been an uneven and unfinished process in which state and capital have explored ways out of the crisis. While consciously leading attempts to reorganize the industry, their efforts have frequently been unco-ordinated, tentative and contradictory. Casualization of labour has given firms greater flexibility. The increasing systemization of design and standardization of components makes production increasingly a process of simple assembly. The detailed, integrated financial controls entailed in 'fast-track' tendering and the reduction in architects' autonomy over design enhance managerial controls over the labour process above and at work-site level, facilitating more effective allocation and co-ordination of labour in the different phases of the work programme.

Employers have also supported the state's sponsorship of narrower-based skills training through the two-year Youth Training Scheme, as opposed to the traditional, broader-based apprentice programme favoured by the unions.[9] But employers continue to express concern about the adverse effects of casualization, especially the industry's poor training record, skill shortages, and its worsening health and safety record. The employers are, moreover, fundamentally concerned that a deregulated labour-market might, in any sustained upturn, expose their excessive dependence on subcontractors as the latter exploit their market power to seek out the most lucrative contracts, and their vulnerability to competitive wage inflation without a co-ordinated bargaining policy. Because in many areas the self-employed are the only labour available, they will be in a position to dictate their own terms. They are not constrained by any collective agreement and their bargaining continues all year round. Government could then face escalating construction costs, adversely affecting its own expenditure plans and any wider economic recovery. Lacking a more immediate means of intervening, it would find it difficult to avoid resorting to the familiar 'stop–go' policy of depressing demand.

Although compared with many factory operatives, construction workers retain a measure of individual control in their work, this is nevertheless increasingly degraded by the subdivision of tasks. The consequences for construction workers' bargaining power has been similarly ambiguous. Some skilled workers have gained from the greater flexibility afforded by firms' casual labour policies, with relatively high earnings, successful avoidance of tax imposts, or the feeling of freedom through not being dependent on any particular 'boss'. These are a minority, however, and many more workers have experienced greater insecurity of employment and earnings.

Our analysis thus differs from both orthodox accounts of the industry stressing its organizational stability (e.g. Fleming, 1977) and explanations of the general economic crisis and transformation of work emphasizing the collapse of Fordism as the dominant mode of capitalist economic regulation (e.g. Piore and Sabel, 1984). Both of these approaches place conditions of demand at the centre of economic restructuring. Although construction does not figure in Piore and Sabel's account of the collapse of Fordism in the massive shift in consumer demand from standardized, mass-produced commodities towards high quality specialized products, their argument closely replicates orthodox explanations of construction's fragmented industrial structure and work organization. Furthermore, there is a similar affinity in their assertions of the intrinsic and extrinsic benefits deriving to labour from up-market specialization. Our analysis suggests that, in construction at least, this argument is implausible.

The development of alternatives to the present atomized system has to take account of the contradictory positions of those involved. Labour-market hierarchies, for instance, conflict with union goals of uniform skills and pay rates. But there is evidence of worker's support for some tighter form of labour-market regulation, and even among the self-employed little overt hostility towards trade unionism (Scase and Goffee, 1981: 743).[10] A major problem for unions, nevertheless, is to make their policies for regulation relevant to workers. Apart from the intrinsic difficulties of organizing a fragmented and mobile workforce, sometimes in the teeth of employer oppositon, the unions have been impeded by a conflict of interest between those who work in the private and public sectors. This division has generated conflicting goals and political tensions among influential power-holders inside the construction unions, for example, between the national leaderships and the local representatives of DLO workers. The conflicts have constrained union proposals for decasualization and public owner-ship, and have also ensured that UCATT and the TGWU, despite the ineffectiveness of NWR 26 and the statutory rules which now prohibit its enforcement by contractors and local authorities, continued their formal opposition to recognizing the self-employed in collective agreements and to bargaining on their behalf. The EETPU (together with the employers' association, the ECA) has recently decided formally to include the self-em-ployed within the regulatory framework for the electrical contracting industry. This is an important development in which the union is clearly attempting to prevent the effective de-unionization of the industry. It represents a concerted shift in formal policy – following more informal practice – towards the regulation as opposed to the exclusion of this form of labour.[11] However, the success of this strategy will be influenced by how employers use the self-employed. Cathcart (1986) has shown, for instance, how depending on size and market position, some firms use them for reasons of numerical flexibility, and others for financial flexibility. It is more likely that joint regulation will be favoured by the former, for whom the labour-only subcontractor is a 'top-up', than by the latter.

As for state regulation, labour-market considerations have always been subordinate to other macroeconomic concerns and remain so. Indeed, the state's role in construction in the 1960s was consistent with its role in the wider economy, which Fine and Harris (1985: 14–18) describe as having been 'directed towards managing aggregate demand' and giving 'inadequate attention to the more directive role of intervening in and guiding capitalist accumulation except on a piece-meal basis'. Asserting a continuity in the 'stop–go' policies associated with balance of payments crises in the 1960s and the deflationary monetarist cuts since 1976, these authors argue that the state failed to provide a sufficient stimulus to restructuring in the absence of planning and a coherent industrial policy. The construction industry illustrates the thesis, and it also shows how the 'institutional structure for intervention has been constructed and reconstructed several times since the war'. This experience suggests both some of the difficulties confronting direct state intervention and some additional influences on the strategies adopted by employers and unions.

As Jessop (1980: 40–1) has convincingly argued – and he might have given the EDC for construction as a prime example – the nature of post-war state planning in Britain has been of a limited, indicative and voluntary kind. Planning was handicapped by the weakness of the state industrial planning apparatus, and by its domination by the Treasury and Bank of England. Dunleavy (1981) has shown how, in the 1960s, central government was hampered by shortages of labour, materials and managerial expertise, and had to oversee the highly devolved construction prog-ramme, which was subject to rising costs and lengthening build times, with only a very small central staff and quite inadequate control mechanisms through the Ministry of Public Building and Works. Meanwhile, the Treasury retained overall control over economic policy and government expenditure, and was able to assert the priority of sterling and the balance of payments over the commitments to growth and employment. An ad-ditional empediment to planning was the fragmented organizational struc-tures and policies of the unions and employers' organizations (see, e.g., Grant and Streeck, 1985: 170).

But institutional weaknesses and rigidities were not the only, or most important, deterrent to restructuring. Any attempt at restructuring the industry would have to take into account the complex structural forces underlying its economic and social organization – its fragmentation, its heterogeneity, and its vital role in the wider economy. Indeed, the con-struction industry's function as a regulator for the rest of the economy has deterred governments from any systematic policy for restructuring. De-spite its economic centrality, however, the industry has few bargaining levers against deflationary policies since investment can be postponed at least in the short term with few adverse effects on other sectors (Dudley, 1983: 114–17; Sugden, 1975: 16). Construction employers have been forced, therefore, to seek the alternative routes out of the crisis and back to profitability discussed above. Their successes contrast sharply with the worsening difficulties facing the unions under a hostile and increasingly strong, centralized state.

Unions generally have been ambivalent towards state intervention, preferring to preserve the basis for 'free collective bargaining' when economic conditions were favourable. Nevertheless, construction unions have been concerned to use the power of the state to decasualize their industry, but they have been hampered by, among other things, their isolation and marginality within the broader labour movement. The logic of their situation had led them to press for policies at a more devolved level of the economy and the industry. Unions found it easier to deal with local government and were often particularly effective in influencing the condi-tions of labour in the devolved public sector construction programme. Indeed, these forms of local state–union regulation have been the target of Conservative government measures cutting back DLOs and introducing legal constraints on the use of contracting power to enforce collective bargaining and union organization.

Apart from these local-level strategies, the unions have continued to argue for the expansion of public ownership and for a statutory scheme for registration of employers and workers. However, even in its most radical form, the former would still leave the bulk of work to be undertaken by the private sector. The latter would have to take account of how, like the scheme for registered dock-workers on which it is modelled, it could avoid being progressively undermined by clients placing their contracts with firms outside the scheme's jurisdiction.

Each of the unions' proposals for restructuring includes items which are logical, expedient, short-term tactical goals. But would they overcome the complex and pervasive character of the industry's problems? As Nolan and O'Donnel (1987) have pointed out, traditional labour movement and recent local-state restructuring strategies share a fundamental contradiction: they contain 'a basic tension between planning *for* and planning *against* the power of the market'. Our analysis has suggested that the potential for planning the industry is limited in the absence of a more comprehensive plan for the economy as a whole.

Notes

1. The analysis is based on the literature and also on our own empirical inquiries into the construction industry, which are part of a wider project (supported by the Leverhulme Trust) on the right to associate in theory and practice. The research has involved over 140 interviews with managers, workers, and representatives of employer's associations and unions in Britain. In addition, the authors wish to acknowledge the assistance of Michael Ball, Jon Cruddas, Peter Nolan, and several persons belonging to employer organizations, firms, local authorities and trade unions who prefer to remain anonymous.
2. Cf. the inter-war years when large firms were better placed to sustain profitability in a recession (Richardson and Aldcroft, 1968: 38)
3. This chapter concentrates on the 'building' trades, though most other sectors of the construction industry are subject to similar trends and developments. Even the electrical contracting sector, which was often quoted as the supreme example of effective industry-wide collective regulation, has recently seen a massive growth of self-employment. The heavy engineering sector is the one remaining major exception to the decline of labour regulation.
4. Financial Times, 4 June 1987, *FT Survey UK Building Industry* p.3.
5. Though not always or for all purposes. For example, in *Ferguson* v. *John Dawson & Partners* [1976] 3A11ER817, a nominally self-employed construction worker was found to be an employee and was therefore able to claim damages from a contractor for breach of statutory duty after an industrial accident. But the finding might have been different if the issue had been one of redundancy or unfair dismissal. The determination for different legal purposes

of employment status – taking into account a wide range of material factors as well as the label attached to the relationship by the parties – has led to a voluminous and contradictory body of case law (Wedderburn, 1986: 110–32; Leighton, 1986).

6. Inducing a firm to break a contract with a labour-only subcontractor was exposed to civil liability in the mid-sixties (*Emerald Construction* v. *Lowthian* [1966] 1WLR 691); since 1980 the legal freedom to engage in industrial action has been severely curtailed.

7. Phelps Brown (1968) was established to investigate the industry's engagement and use of labour 'with particular reference to labour-only subcontracting'. For a critique of its recommendation, see Lewis (1969). The issue of registration was subsequently explored by Forbes (1972) and self-employment by the abortive Misselbrook inquiry, 1972–4. The major investigations of strikes were Cameron (1967), NEDO (1970) and CIR (1972a). In the late 1960s a series of reports on pay bargaining was published by the NBPI (1968a,b,c,d and 1969). See too CIR (1972b).

8. ˆHC Bill 1969–70 [147].

9. Cf. the TGWU's proposal that training for the whole industry should be revamped with the CITB acting as managing agents allocating trainees to particular firms: House of Commons Select Committee on Employment 1986–7, *Skills Shortages*, Minutes of Evidence, 197, 4 March 1987, pp. 69–75. Employers' organizations are divided over whether responsibility for training should be retained by individual firms or whether, as the FMB recently advocated, the CITB should co-ordinate and control all the training needs and machinery of individual trades and crafts: Construction Industry Training Board, *More Effort Needed to Train Subcontract Building Workers*, PR 638, 16 November 1987.

10. In electrical contracting, for instance, a former employment agency recently claimed to have obtained registration from the Certification Officer as an independent 'Union of Self-Employed Electricians' with 3,000 members, and intends to bargain with contractors. There, too, is a growing tendency for the self-employed to continue purchasing JIB Benefit stamps, through their 'employer', which give holiday pay, private medical care, and death and injury benefits.

11. JIB Industrial Determination, National Working Rule 18, *Labour-Only Subcontracting and Temporary Use of Self-employed Labour*, December 1987. Similar changes are envisaged in building, where the BEC has tabled proposals, now under consideration by the unions, to amend NWR 26 to extend coverage to the self-employed and allow them to be 'engaged under terms and conditions which, taken as a whole, are not less favourable to the operatives' i.e. direct employees.

Bibliography

Allen, V.L. 1952. 'Incentives in the Building Industry', *Economic Journal*. Vol. 62, 595–608.

Austrin, T. 1980. 'The "Lump" in the UK Construction Industry', *Capital and Labour: A Marxist Primer*. Ed. T. Nichols. Glasgow: Fontana, 300–13.

Ball, Michael. 1980. 'The Contracting System in the Construction Industry'. Birkbeck College Discussion Paper No. 86. London: Department of Economics, Birkbeck College.

—— 1983. *Housing Policy and Economic Power: The Political Economy of Owner Occupation*. London: Methuen.

—— 1988. *Rebuilding Construction: Economic Change in the British Construction Industry*. London: Routledge & Kegan Paul.

Bishop, Donald. 1975. 'Productivity in the Construction Industry', *Aspects of the Economics of Construction*. Ed. D.A. Turin. London: George Godwin, 58–96.

Bresnen, M.J., K. Wray, A. Bryman, A.D. Beardsworth, J.R. Ford and E.T. Keil. 1985. 'The Flexibility of Recruitment in the Construction Industry: Formalisation or Re-Casualisation?', *Sociology*. Vol. 19, No. 1, 108–24.

Burgess, Keith. 1975. *The Origins of British Industrial Relations*. London: Croom Helm.

Cameron. 1967. Court of Inquiry into Trade Disputes at the Barbican and Horseferry Road Construction Sites in London. *Report*. Cmnd 3396. London: HMSO

Carter, C.F. 1958. 'The Building Industry', *The Structure of British Industry*. Ed. D. Burn. Cambridge: Cambridge University Press, 47–75.

Cathcart, Robert L. 1986. 'Labour Only Subcontracting and Flexible Working Practices in the Electrical Contracting Sector of the Construction Industry: A Study of the Incidence of Self-Employment and Flexible Working in the Greater London Region'. MA dissertation, University of Warwick.

Centre for Construction Market Information (CCMI). 1985. *Survey on Management Contracting*. London: CCMI.

Channon, D.F. 1978. *The Service Industries*. London: Macmillan.

Chesterman, M. 1982. *Small Businesses*. London: Sweet & Maxwell.

Clarke, Linda. 1980. 'Subcontracting in the Building Industry'. *Production of the Built Environment*. Proceedings of the Second Bartlett Summer School. London: University College, 35–52.

Commission on Industrial Relations (CIR). 1972a. *Employers' Organisations and Industrial Relations*. CIR Study 1. London: HMSO.

—— 1972b. *Alcan Smelter Site*. Report No. 29. London: HMSO.

Davies, P. and M. Freedland. 1984. *Labour Law: Text and Materials*. 2nd edn. London: Weidenfeld & Nicolson.

Department of the Environment (DoE). 1986. *Housing and Construction Statistics 1975–1985 Great Britain*. London: HMSO.

Druker, Janet. 1980. 'One Big Union? Structural Change in the Building Trade Unionism'. PhD thesis, University of Warwick.

Dudley, G. 1983. 'The Road Lobby: A Declining Force?', *Pressure Politics*. Ed. D. Marsh. London: Junction Books, 104–28.

Dunleavy, Patrick. 1981. *The Politics of Mass Housing in Britain, 1945–1975*. Oxford: Clarendon Press.

Durcan, J.W., W.E.J. McCarthy and C.P. Redman. 1983. *Strikes in Post-War Britain: A Study of Stoppages of Work Due to Industrial Disputes, 1946–73*. London: Allen & Unwin.

Economist Intelligence Unit (EIU). 1981. *Capital Spending and the UK Economy*. London: EIU.

Fine, B. and L. Harris. 1985 *The Peculiarities of the British Economy*. London: Lawrence & Wishart.

Flanagan, R., G. Norman, V. Ireland and R. Ormerod. 1986. *A Fresh Look at the U.K. & U.S. Building Industries*. Report prepared for the Building Employers Confederation by the Department of Construction Management, University of Reading. London: Building Employers Confederation.

Fleming, M.C. 1977. 'The Bogey of Fragmentation in the Construction Industry'. *National Builder*, Vol. 58, 134–7 and 284–6.

—— 1980. 'Construction', *The Structure of British Industry*, Ed. P.S. Johnson. St Albans: Granada, 231–53.

Forbes, Sir Hugh. 1972. *Enquiry into the Registration of Builders*. London: HMSO.

Gagg, Max. 1969. 'The Subby Bricklayer', *Work 2*. Ed. R. Fraser. Harmondsworth: Penguin, 130–46.

Gallagher, T.J. 1984. *Industrial Relations on Site*. London: Construction Press.

Grant, Wyn. 1983. 'The Organisation of Business Interests in the UK Construction Industry'. Discussion Paper. Berlin: International Institute of Management.

—— and W. Streeck. 1985. 'Large Firms and the Representation of Business Interests in the UK and West German Construction Industry', *Organised Interests and the State: Studies in Meso-Corporatism*. Ed. A. Cawson. London: Sage, 145–73.

Health and Safety Executive (HSE). 1987. *Report by HM Chief Inspector of Factories 1986–87*. London: HMSO.

Hillebrandt, Patricia M. 1984. *Analysis of the British Construction Industry*. London: Macmillan.

Hilton, W.S. 1968. *Industrial Relations in Construction*. London: Pergamon.

Institute of Personnel Management/Incomes Data Services (IPM/IDS). 1986. *Competitive Tendering in the Public Sector*. London: IPM/IDS.

Jessop, Bob. 1980. 'The Transformation of the State in Post-War Britain', *The State in Western Europe*. Ed. R. Scase. London: Croom Helm, 23–93.

Karn, Valerie. 1985. 'Housing', *Between Centre and Locality: The Politics of Public Policy*. Ed. S. Ranson, G. Jones and K. Walsh. London: Allen & Unwin, 163–85.

Lange, Julian E. and Daniel Quinn Mills. 1979. *The Construction Industry: Balance Wheel of the Economy*. Lexington, MA: Lexington.

Langford, D.A. 1982. *Direct Labour Organisations in the Construction Industry*. Aldershot: Gower.

Layton, John. 1985. 'How Do DLOs Measure Up to Companies?, *Public Finance and Accountancy*. 13 September, 10–13.

Leighton, Patricia. 1986. 'Marginal Workers', *Labour Law in Britain*. Ed. Roy Lewis. Oxford: Blackwell, 503–27.

Lewis, Richard. 1982. 'Contracts between Businessmen: Reform of the Law of Firm Offers and an Empirical Study of Tendering Practices in the Building Industry', *Journal of Law and Society*. Vol. 9, Winter, 153–75.

Lewis, Roy. 1969. 'Report of the Phelps Brown Committee', *Modern Law Review*. Vol. 32, January, 75–80.

—— and R. Simpson. 1986. 'The Right to Associate', *Labour Law in Britain*. Ed. Roy Lewis. Oxford: Blackwell, 47–79.

Loughlin, Martin. 1986. *Local Government in the Modern State*. London: Sweet & Maxwell.

McCarthy, W.E.J. 1964. *The Closed Shop in Britain*. Oxford: Blackwell.

McCormick, B.J. 1980. 'Trade Union Reaction to Technological Change in the Construction Industry', *Industrial Relations and the Wider Society (Aspects of*

Interaction). Ed. B. Barrett, E. Rhodes and J. Beishon. Milton Keynes: Open University Press, 171–89.

McGhie, B. 1982. 'The Implications of Project Management', *Production of the Built Environment*. Proceedings of the Third Bartlett Summer School. London: University College, 2.1–9.

McKinlay, Alan. 1986. 'The Management of Diversity: Organizational Change in the British Construction Industry, c. 1960–1986'. Working Paper. Birmingham: Work Organization Research Centre, Aston University.

Marsh, A., P. Heady and J. Matheson. 1980. *Labour in the Construction Industry: Final Report to the CIMB*. London: Department of the Environment.

Ministry of Public Building and Works. 1964. *The Placing and Management of Contracts for Building and Civil Engineering Work*. London: HMSO.

Ministry of Works. 1944. *the Placing and Management of Building Contracts: Report of the Central Council for Works and Buildings to the Minister of Works*. London: HMSO.

Morton, C.N. 1979. 'Collective Bargaining in Building and Civil Engineering: A Case Study of Three Major Re-development Projects in the City of London'. PhD thesis, University of London.

National Board for Prices and Incomes (NBPI). 1968a. *Report on a Settlement Relating to the Pay of Certain Workers Employed in the Thermal Insulation Contracting Industry*. Report No. 84. Cmnd 3784. London: HMSO.

—— 1968b. *Pay and Conditions in the Civil Engineering Industry*. Report No. 91. Cmnd 3836. London: HMSO.

—— 1968c. *Pay and Conditions in the Building Industry*. Report No. 92. Cmnd 3837. London: HMSO.

—— 1968d. *Pay and Conditions in the Construction Industry Other than Building and Civil Engineering*. Report No. 93. Cmnd 3838. London: HMSO.

—— 1969. *Pay and Conditions in the Electrical Contracting Industry*. Report No. 120. Cmnd 4097. London: HMSO.

National Economic Development Office (NEDO). 1970. *Large Industrial Sites: Report of the Working Party on Large Industrial Construction Sites*. London: HMSO.

—— 1975. *The Public Client and the Construction Industries: Report of the Building and Civil Engineering Economic Development Committees' Joint Working Party Studying Public Sector Purchasing*. (Wood Report). London: HMSO.

——, Building Economic Development Committee. 1983. *Faster Building for Industry*. London: HMSO.

——, Building and Civil Engineering Economic Development Committees. 1986. *Construction Forecasts 1986/87/88*. June. London: HMSO.

Nolan, P. and K. O'Donnell. 1987. 'Taming the Market Economy: A Critical Assessment of the GLC's Experiment in Restructuring for Labour', *Cambridge Journal of Economics*. Vol. 11, September, 251–64.

Phelps Brown. 1968. *Report of the Committee of Inquiry under Professor E.H. Phelps Brown into Certain Matters Concerning Labour in Building and Civil Engineering*. Cmnd 3714. London: HMSO.

Pinschof, Maria. 1970. *Men on Site: Ten Case Studies in Building Management*. London: HMSO.

Piore, M. and C. Sabel. 1984. *The Second Industrial Divide: Possibilities of Prosperity*. New York: Basic Books.

Price, Richard. 1980. *Masters, Unions and Men: Work Control in Building and the Rise of Labour 1830–1914*. Cambridge: Cambridge University Press.

Ramsdale, Phillip. 1985. 'Evaluating the DLO Legislation'. *Public Finance and Accountancy*, 13 September, 6–10.

Routh, G. 1980. *Occupation and Pay in Great Britain 1906–79*. 2nd edn. London: Macmillan.

Richardson, H.W. and D.H. Aldcroft. 1968. *Building in the British Economy between the Wars*. London: Allen & Unwin.

Scase, R. and R. Goffee. 1981. '"Traditional" Petty Bourgeois Attitudes: The Case of Self-Employed Craftsmen', *Sociological Review*. Vol. 29, No. 4, 729–47.

Sugden, J.D. 1975. ' The Place of Construction in the Economy', *Aspects of the Economics of Construction*. Ed. D.A. Turin. London: George Godwin, 1–24.

Turner, Dennis. 1987. 'The Construction Industry in Britain', *Midland Bank Review*. Autumn, 16–23.

Uff, J. 1985. *Construction Law*. 4th edn. London: Sweet & Maxwell.

Undy, R., V. Ellis, W.E.J. McCarthy and A.M. Halmos. 1981. *Change in Trade Unions: The Development of UK Unions since the 1960s*. London: Hutchinson.

Union of Construction, Allied Trades and Technicians (UCATT). 1986. 'The State of the Trade of the Industry', *UCATT Information Bulletin*. No. 18 (Spring), 1–6.

Wedderburn, K.W. (*Lord*). 1986. *The Worker and the Law*. 3rd edn. Harmondsworth: Penguin.

Winch, G.M. 1985. 'The Construction Process and the Contracting System: A Transaction Cost Approach', *Production of the Built Environment*. Proceedings of the Seventh Bartlett Summer School. London: University College, 262–70.

—— 1986. 'The Labour Process and Labour Market in Construction', *International Journal of Sociology and Social Policy*. Vol. 6, No. 2, 103–16.

Textiles

The previous studies have concerned the failure of governments to promote the reorganization of highly fragmented industries. The company in this study has benefited greatly from government promotion of concentration in the textile industry, enabling it to develop a flexible strategy to cope with intense international competition.

The textile industry has been associated with attempts to bolster profitability through the export of capital to low-wage countries. This study criticizes those theories which emphasize such a response over the more diverse strategies available to multinationals. It emphasizes how changes in the nature of competition in both the textile and clothing sectors have promoted responses to technical change and investment in the advanced economies to complement the export of capital. 'Upmarket restructuring' in the case of this company has been facilitated by its dominant position in the domestic market.

For the firm, this has been a successful strategy. The benefits to workers and the community often claimed for this form of restructuring, are, however, challenged in this study. It shows how the investment decisions of the company have not resulted in increased task variety but have further deskilled already routinized work processes. Supporting wider evidence that union co-operation with new technology is more often the case than the exception, it illustrates how this co-operation between management and labour has been promoted by craft and gender divisions.

4

Capital Restructuring and Technological Change: A Case Study of a British Textile Multinational

Janet Walsh

Introduction

This chapter highlights the complexities of capital restructuring in the British textile industry by focusing in detail on the strategies and practices of a major British multinational. In response to heightened competitive pressures, the company is attempting to restore the profitability of its textile group by introducing new technology. The recent re-equipment of one of its spinning mills neatly encapsulates these developments and provides the substance of the following discussion. The analysis situates this type of restructuring in the context of broader international patterns of industrial change and development within the industry. The implications for social relationships at workplace level are then analysed, with particular emphasis on the nature of the impact of new technology on work processes and job content, and the formulation of strategies and responses by workers and their union to these developments.[1]

The Changing Nature of the Textile Industry

The theme of industrial change in the textile and clothing industries has generated a considerable amount of academic literature over the last decade. Two distinct perspectives mentioned in the introductory chapter –

the 'new international division of labour' and 'flexible specialization' theories – have drawn upon these complex developments to illustrate general arguments relating to the nature of contemporary capitalist restructuring. These empirical patterns of change and development will therefore be outlined and analysed to show how they have informed both theories. This will enable a more critical discussion of the partiality and one-dimensional nature of these perspectives. It will be argued that the forces which induced one particular textile manufacturer to reorganize substantially the technical nature of the production process at plant level can be more adequately seen as the product of a set of historically specific relationships.

Empirical Patterns of Change
There can be no doubt that an international restructuring of production and employment in the textile and clothing industries has occurred. In the period 1963–80, the share of OECD countries in world production of textiles and clothing fell substantially (OECD, 1983:12), with developing countries and centrally planned economies accounting for a greater share of world output. Employment, too, has been falling since 1963 in textiles in the OECD area and since 1973 in clothing, while there has been continued employment growth in developing countries and centrally planned economies. In terms of the European Community, the textile industry was among the largest shedders of labour in a period when manufacturing employment was falling (Shepherd 1981:6–7) and the overall unemployment rate was rapidly rising.

The decline of the cotton textile industry in most Western countries has been particularly dramatic (Cable and Baker, 1983:45–6). Cotton yarn production in the UK has fallen to under a third of 1967 levels and production in the USA has been halved. Substantial reductions have also been reported in France, West Germany and Japan. This has been associated, during the 1960–80 period, with a seven-fold increase in textile production in South Korea and Thailand respectively. In short, the evidence seems to suggest that this industry has undergone a substantial internationalization of its productive activities, characterized by the relocation of production to lower waged developing countries.

Nevertheless, recent empirical investigations concerned with the clothing industry (Mitter 1986:49; Zeitlin 1985:5) have argued that complex changes are occurring in the nature of global competition, which tend to favour the garment manufacturers of developed countries. This has been attributed to a multitude of factors, including changes in consumer tastes, demographic changes in the structure of the population, the volatile nature of demand for fashion, and aggressive product differentiation by the industry, which have fragmented the mass market for clothing and reduced the economic viability of long-run manufacture. In response to intensified competition, speedy response to fashion changes has become a major strategy of clothing retailers and productive flexibility is therefore required at all stages of the garment manufacturing process often involving the use of micro-electronic technology. Consequently, hoarding labour on the factory floor has become unnecessary, and outside sourcing (subcontracting and homeworking) is now proliferating. Such developments are exemplified by the Italian clothing firm, Benetton, which has its garments made by 11,500 workers, only 1,500 of whom work directly for Benetton. The majority are employed by subcontractors in small factories, of 30–50 workers, each regulated by a centralized, high technology-oriented design, sales and marketing operation (cf. Murray, 1985). The ability of garment manufacturers to react to short-run changes in fashion is also partly dependent on geographical factors, particularly the proximity of such manufacturers to the retailers and ultimately the markets of developed economies.

The complex reconfiguration of the clothing industry has resulted in substantial competitive advantages accruing to the domestic manufacturers of developed countries. Furthermore, changes in the nature of global competition in the garment industry are relevant to any discussion of developments in the British textile industry. Both industries have a close, almost symbiotic relationship.

According to the OECD (1983:29) clothing accounts for roughly 50 per cent of total final fibre consumption in OECD countries, thereby constituting the major determinant of demand for textile products. It is hardly surprising to discover that a conglomeration of symmetrical pressures is leading to a restructuring of textile production in a direction which potentially favours domestic producers who can react flexibly to swift changes in fashion requirements, whilst offering high quality output embodying sophisticated design and colour co-ordination.[2] Again, this tends to have meshed with a long-run trend of the textile industry towards increasingly capital-intensive methods of production. Cable and Baker (1983:30,57) have noted that in the spinning and weaving sectors, technology has provided individual firms in relatively high labour cost locations with the possibility of resisting competition from developing countries by virtue of enormous productivity increases and a dramatic reduction in unit costs of production. It is therefore clear that the textile industry exhibits complex and diverse patterns of restructuring and not single, unilinear trends.

Theoretical Perspectives
Both 'new international division of labour' and 'flexible specialization' theories draw selectively upon these complex developments to highlight one dominant form of capitalist restructuring. New international division of labour' theorists have used the example of the internationalization of textile and clothing production from the late 1960s onwards to argue that the central response of capitalists to profitability crisis has been to relocate to lower waged countries. Clairmonte and Cavanagh have argued that the

textile industry is 'one of the leading industries in the changing international division of labour' and is therefore the 'major battleground within and between the underdeveloped capitalist economies and the developed capitalist economies' (1981:165). Fröbel *et al.* (1980) have argued that the internationalization of productive activity was a classic illustration of fundamental changes characterizing the manufacturing industries of advanced capitalist countries, and represented a new stage of imperialist development undermining the classical international division of labour which was based on the integration of developing countries into the world economy as raw material producers.

Although many developing countries have their own indigenous manufacturing industries, Fröbel *et al.* (1980) focused on the tendency of multinational corporations to relocate labour processes in the lower waged countries of the Third World. Here, competition between workers faced with a vast industrial reserve army enables capital to impose low wages, long hours and high work rates in poor conditions. Typically, production occurs in state-sponsored duty free enclaves or free trade zones for the world market. Conversely, the crisis of profitability in advanced capitalist countries is understood as the result of rising labour strength throughout the post-war period, which has made it difficult to maintain the profit rate through increasing the rate of exploitation of labour (cf. Arrighi, 1978; Landsberg, 1979). The strengthening of the working class is a direct result of its concentration, which accompanies the concentration of capital (Arrighi, 1978) or the result of the post-war boom which largely eliminated the industrial reserve army in developed countries.

This type of theorization of crisis has been disputed, however, by Fine and Harris (1979) and Jenkins (1984). They argue that the falling rate of profit, capitalist crisis and the internationalization of capital can be more adequately conceptualised in terms of Marx's analysis of the circuits of capital, and not merely exhange-based categories or distributional relationships.

An exclusive or one-sided emphasis on the geographical relocation of multinational capital to lower waged developing economies can seriously underestimate the dynamic and complex nature of capitalist responses to profitability crisis. The problem with 'new international division of labour' theories is that they overemphasize one pattern of capital restructuring – the relocation of production in pursuit of 'cheap labour' – and therefore neglect powerful counter tendencies, particularly the ability of firms in the developed economies to use technology to reduce unit costs of production. As Jenkins (1984) argues, reducing labour costs by the fragmentation and relocation of production processes is only one possible strategy available to firms, and this is unlikely to be the dominant stategy.

Theories of restructuring based on notions of a new model of competitive behaviour characterized by 'flexible specialization' or 'neo-Fordism' accord quite plausibly with more contemporary developments in the textile industry. Here, the crisis is seen as a period of evolution from Fordist methods of production, involving mass production of basic, standardized products for a homogeneous, undefined set of consumers to an emerging stage of neo-Fordism characterized by small or medium batch production in specialized, high quality goods for discrete categories or segments of consumers. This 'upmarket' restructuring of production has meshed with the development of microelectronic technology, particularly computerized design and production technology, which enables small batches to be produced at low cost.

The problematic aspects of this thesis have already been mentioned in the introductory chapter, particularly the emphasis 'flexible specialization' theorists place on the fragmentation of consumer demand as the cause of these developments and the extent to which it is indicative of any more than a minority of manufacturing industries. However, it is the implication that this pattern of restructuring can yield positive benefits for labour, in the form of skill enhancement and improved autonomy, that demands particular investigation (cf. Piore and Sabel, 1984; Streeck, 1986, Zeitlin, 1985). Alternatively, it is also a well-established phenomenon that new technology can be used to deskill and displace labour. This case study permits an investigation of the impact that new technology can have on work processes and skill requirements, and the implications for labour of such developments.

Forces for Change

In the following discussion, capital restructuring and its impact at work-place level will be analysed by reference to several sets of mediating relationships. These are the relationship between the state and the textile industry, the nature of intercapitalist competition, relationships between workers at the level of the plant and the overall relationship between capital and labour.

The historical context of state intervention in the industry, and the partial and incomplete nature of such initiatives provides the background to the contemporary restructuring of the company's textile group and the modernization and re-equipment of one of its spinning mills. The nature and degree of intercapitalist competition has an important bearing on the type of strategies adopted by individual firms in response to profitability crises. Although in the instance of this case study, competition has led to the implementation of the most advanced techniques of production, in the nineteenth-century cotton textile industry it tended to divide capitalists by restraining them as individuals from engaging in restructuring the technical nature of production (Lazonick, 1979).

Relationships within the working class are also highly significant for an understanding of restructuring. Relationships between male and female workers, different ethnic groups and groups of workers with different skill, can play a crucial role in determining the structure of the division of labour which emerges from technical change, and the nature and rapidity of the implementation of technical innovations. These relationships mediate the overall capital–labour relation, which constitutes the terrain upon which employers and workers co-operate or conflict in the implementation of restructuring strategies.

Contemporary patterns of restructuring in the industry will be analysed through the medium of these relationships; the historical background of state intervention in the industry, intercapitalist competition, relationships between workers and the overall capital–labour relation. Particular attention will be focused on the ways in which, in the current period, they have provided a positive context for radical changes in the technical structure of production in the company concerned. This study, however, is not untypical of broader patterns of change across a range of OECD countries. Therefore, to situate developments coherently within the UK industry, the structural changes characterizing the textile industries of developed economies will be briefly explored.

International Patterns of Restructuring in the Textile Industry

The relocation of production to lower waged developing countries has been one strategy pursued by a number of textile firms to restore and enhance profitability. Nevertheless, there is evidence to suggest that this is only one way that capital can restructure production and that its significance for the textile industries of developed economies is being eclipsed by other factors.

Although an OECD study (1983:57) has highlighted the substantial internationalization of textile and clothing production, relocation may take a number of forms. Interestingly, only relatively limited use of foreign direct investment has occurred over the last 10–15 years, and the bulk of it was aimed at other developed economies. The major form internationalization has taken is 'offshore processing' which constitutes 'an arrangement through which individual production processes are transferred abroad on the initiative of the supplier of inputs who is also the buyer of the transformed product' (p. 58). If it is carried out by an independent unit this is termed subcontracting, which guarantees both the supply of the input and the disposal of the product. This is a means of competition between producers in high cost areas whereby the profits earned as low priced items are used to offset losses on domestic, high quality output. This has increased in recent years but its relative importance varies from country to country. For example, it is an established trend in West Germany, America, Sweden and the Netherlands, but is relatively rare in the UK, Italy and Japan (p. 61). Nevertheless, Toyne *et al.* (1984:142), in a series of interviews with one French and two West German textile firms, noted some disenchantment with offshore processing because it erodes the control that management has over the production process. These firms believed close control required the physical proximity of management and plant which would generate greater flexibility and higher rates of productivity, thereby offsetting the advantage of lower wage rates from offshore processing.

The need to establish productive flexibility in response to intercapitalist competition has precipitated changes in industrial structure. During the post-war period, the structure of the textile industry was characterized by a large number of family firms. But state support for industrial restructuring, to promote productivity improvements and rationalization to make the industry more competitive in the face of low-cost imports and overcapacity, has combined with the tendency of large synthetic fibre producers to engage in strategies of vertical integration. This has generated high levels of concentration in textiles in the UK, France, Japan and the USA, albeit with the important exceptions of Germany and Italy. The OECD study points out, however, that 'since the early 1970s, the movement of concentration has run into strong opposite currents, both in textiles and in clothing' (1983:27), and concludes that there has been a generalized but sharp reduction in the number of enterprises but without any evident trend towards increased concentration. The industrial structure is undergoing change in the direction of more divergent, decentralized corporate structures to maximize flexibility, although this does not imply that ownership is becoming fragmented. In any case, these developments seem to be compatible with a higher level of concentration in basic textiles (spinning, weaving) than in the rest of the industry.

In response to intensified international competition, the pace of technical change has greatly accelerated in the post-war period. The OECD (1983:18) identifies the stimulus to such innovations as the need to economize on unit costs of production to compete more effectively with lower waged economies. Firms in developed economies have opted for automated labour saving technologies – rotor spinning and shuttleless weaving – and Western Europe in general has engaged in scrapping redundant machinery (42 per cent of 1973 weaving capacity and 23 per cent of ring spindle capacity was scrapped in 1973–8). In short, technical change has provided individual firms in relatively high labour costs locations with the possibility of resisting competition from developing countries by substituting capital for labour, raising productivity and dramatically lowering unit labour costs. Such increases in labour productivity have a profound effect on employment levels, more so than imports. Cable (1977:41) has analysed employment changes in clothing, cotton textiles, yarns and footwear in the UK between 1970–5, and concludes that productivity growth emerged as twice as important as trade factors in job displacement. A more recent analysis concerning the impact of technological change on employment in the textile industry (Soete, 1984:125–73) took the longer time scale of 1954–1981 and concluded that although trade factors were quite significant and amounted to 4,900 job losses a year, this was marginal compared to the combined effect of a general slowdown in demand and productivity increases leading to 17,288 job losses a year (Soete, 1984:159–68). Furthermore, over the period 1970–9, the import job displacement effect in textiles originated primarily from EEC countries and not the Third World. Again, such an effect was relatively small compared to the impact on employment of falling demand and rapid productivity increases.[3]

In any case, according to *Textile Horizons* (1986:12), hourly labour costs are becoming less important in affecting the profitability of the industry. In the basic textile industry, labour costs can be limited to between 8 per cent and 20 per cent of sales turnover in modern American and European plants, due to the capital-intensive nature of the industry. The Cable and Baker study for the Economist Intelligence Unit (1983:57) states that in 'textiles, therefore, low labour cost countries are unlikely to have a strong trading advantage.' Gibson (1984:36) has also pointed out that as automation displaces labour 'the competitive edge will stay with the firm that can quickly muster high grade technical and electronic specialists help to maintain sophisticated equipment, and that paradoxically such help may well cost more in a low wage cost country.' Moreover, increasing quality consciousness, rapid fashion changes, the creation of new designs and materials means that an increasing share of output consists of products that do not directly compete with Third World output. Hence, for textile and clothing firms in developed countries, there is a need for proximity to centres of demand because of the dual requirements of flexibility and high quality. This again overrides the disadvantage of relatively higher labour costs in these countries.

The British Context

Restructuring the Textile Industry and the Role of the State – An Historical Overview

Present changes occurring within the British textile industry and the strategies pursued by leading producers to restore and enhance profitability, have been shaped in profound and significant ways by the industry's historical development. Thus, while the industry's fragmented structure has figured prominently in historical accounts of the sector's relative decline, textile production in Britain is now highly concentrated and dominated by three large corporate groups. The attempts by textile firms and the state to reorganize and modernize the industry form the focus of the following discussion. The reasons why corporate strategies emphasizing mass market production based on a high degree of vertical integration and supported by partial state initiatives, failed to guarantee the industry's long-term competitiveness are then illustrated by analysing recent events in one company.

It has been the industry's *inability* to restructure its productive base and to invest in new technology that has provided the focus for much historical research. Lazonick (1979) notes that, in the highly fragmented cotton textile industry of the nineteenth century, manufacturers failed to introduce new technology in the form of the ring frame. Instead, efforts were made to intensify labour with existing technology due to the persistence of an internal subcontract system. This mode of labour of management, which lasted until the demise of mule spinning in the 1960s, involved chief spinning operatives or minders directly employing and paying their own assistants or piecers. Piecers' wages were set in accordance with a negotiated wage list and did not vary with their productivity, whereas minders were paid by the piece. Hence, when normal production levels were exceeded because of intensification, the minders benefited materially, to the exclusion of the piecers who worked longer and harder without additional remuneration. Attempts by individual firms to reorganize this hierarchical division of labour were limited by the highly competitive conditions prevailing in the cotton industry, in which an initial fall in productivity could spell doom for an innovating firm even though over the long run it may become more productive.

The fragmentation of the industry's structure was to remain a problem until the 1960s, notwithstanding the various attempts by the state to promote reorganization and large-scale production economies. As early as 1913, it was clear that the cotton industry in Lancashire was facing acute problems posed by the erosion of Britain's export markets. This was due to the emergence of competition from new exporters, including Japan, and the development of indigenous textile production by newly industrializing countries, notably India. A consequence of the decline in export markets was the emergence of substantial excess capacity due to the existence of a large number of relatively small, horizontally specialized firms which had evolved to cater for large foreign markets. For example, Miles (1976:185) estimates that in 1930 only 58 per cent of spinning capacity and 54 per cent of weaving capacity was in use. Surplus productive capacity in turn discouraged investment in new equipment and raised costs of production.

Initial attempts at restructuring involved rationalization by amalgamations in spinning and weaving. Nevertheless, the industry's capacity was still too large in relation to its existing markets and this provoked varying degrees of state intervention. An early but unsuccessful form of state-initiated restructuring was the Cotton Spinning Industry Act of 1936 which aimed to eliminate redundant spinning machinery. A mere 13 per cent of 1935 installed capacity was eventually scrapped and two public inquiries conducted during and immediately after the Second World War[4], emphasized the necessity for investment in new machinery and radical changes in industrial structure.

In the immediate post-war period, state policy in the textile sector was more *dirigiste* and consisted of financial subsidies to promote rationalization and re-equipment . This culminated in an ambitious piece of sectoral intervention – the 1959 Cotton Industry Act. In the context of a situation in which only two thirds of spindles and looms were operational in 1958, the Act proposed that half of the industry's plant be scrapped in spinning and 40 per cent in weaving and finishing. Firms were compensated for their losses by the provision of grants to those ceasing to trade in textiles and subsidies for the purchase of new equipment. A total of £17 million was paid in compensation, £11.3 million of which was contributed by the state. According to Miles (1976:192) the rationalization target was almost reached, with 48.1 per cent of spinning spindles and 38 per cent of looms scrapped, and 203 firms left the industry (one fifth of the total). Nevertheless, the problem of excess capacity still remained because only relatively worthless, antiquated machinery was scrapped[5] and re-equipment, too, fell short of government expenditure estimates.[6] This may be attributable to general uncertainty about the industry's long-term future in the context of rising imports[7] (cf. Miles 1976:191). The Act also left untouched the basic problem of the industry's structure, with a large number of small firms still engaging in horizontal specialization.

Contemporary government assessments deemed this type of sectoral intervention a failure. During the period 1961–2, the Estimates Committee of the House of Commons argued that 'failing a speedy and satisfactory solution to the related problems of imports, marketing and the fuller use of plant and machinery, much of the expenditure incurred would have been to no purpose' (cited in Miles, 1976:192). Miles's (1976) more recent assessment of the 1959 Act was that it made little impact because small companies were unable to take advantage of financial incentives due to lack of a capital base, poor cash flow and inadequate managerial expertise.

The apparent failure of the 1959 measures to enhance the international competitiveness of the cotton textile industry meant that, from the early 1960s, the state adopted a different strategy based less on financial assistance and subsidy, and more on protectionist measures, and the encouragement of merger activity to generate capital concentration and vertical integration.

The most sustained impact on the industrial structure of the textile industry came from the acquisitions pursued by the two leading synthetic fibre producers in the 1960s. Both companies, Courtaulds and ICI, needed a secure customer base for their synthetic fibre products, and fear for the future viability of their domestic markets motivated both companies to intervene in the textile industry. Courtaulds, in particular, acquired a substantial share of the UK textile industry and became active in a wide range of textile processes ranging from spinning to clothing manufacture. Consequently, between 1959–67 the total number of firms in the industry fell by 36 per cent, and there was a marked shift to a more concentrated, vertically integrated industrial structure.

This strategy was encouraged by the state. According to Sir Arthur Knight (Chairman of Courtaulds 1975–9), 'the initiative in approaching the fibre producers came from the government side. It was made clear that the government could not be expected to do more to help the industry, financially or by seeking agreement with exporting countries, unless the industry were seen to be doing more to help itself. Courtaulds felt encouraged by Board of Trade officials to take an initiative. The objective was to form a strong group which could effectively manage the troubled Lancashire section of the industry' (1974:52). Thenceforth, Courtaulds was to play a pivotal role in the industry and helped shape future state policy towards the textile sector.[8] In particular, the company was an influential exponent of the argument that modernization of the industry necessitated a period of special protection from low-cost imports.

The industry eventually got the protection it sought, in a new comprehensive quota system for cotton textiles introduced in 1966 covering all newly and semi-industrialized countries excluding the Sino-Soviet bloc. During the early 1970s, this was subsequently replaced by a system which allocated specific quotas to individual countries and during 1972, tariffs were introduced. In addition, Britain became a signatory to international protectionist agreements.[9]

Up to 1964, the Conservative government displayed few public doubts about the extent and effect of acquisitions by the synthetic fibre producers but the Labour government adopted a more ambivalent stance. Despite the positive attitude to mergers adopted by the Industrial Reorganisation Corporation, Courtaulds' monopoly of the supply of man-made cellulosic fibres was referred to the Monopolies Commission for investigation in 1965. While the Commission was sitting, the company continued to acquire firms but the clearance for such acquisitions was being delayed by the Board of Trade. According to Knight (1974:153), 'this led to some irritation because Courtaulds were being encouraged to get on with integration, by both the Prime Minister in private conversation and the President of the Board of Trade (Douglas Jay), and the punctiliousness of the Board's officials contrasted oddly with this pressure.' Moreover, in 1968 the Commission concluded that Courtaulds' acquisition of firms in the textile industry was much greater than required 'to develop and promote the use of its [the company's] fibres and keep it in touch with users' problems' (1968:85). It also announced that the company should not make any further acquisitions in any sector of the textile and clothing industries, if its share of capacity (or of sales) exceeded 25 per cent. Following Courtaulds' takeover bid for English Calico (now Tootal) in 1969, the government announced a prohibition on further mergers between any of the five largest companies in the textile industry (such firms, however, could continue to acquire smaller firms in the industry). Nevertheless, this moratorium was breached when ICI made a takeover bid for Viyella International, to effect a merger with Carrington and Dewhurst, although ICI had to reduce its eventual shareholding in Carrington Viyella to not more than 35 per cent.

Meanwhile, the state reoriented its attention to smaller and medium-sized textile firms to act as a counterweight to the influence of the larger corporate groupings. In 1970, the Industrial Reorganisation Corporation administered a special loan fund for medium-sized and smaller textile companies to finance re-equipment and expansion schemes and up to £10 million was earmarked for textiles. Textile companies also benefited from state financial assistance in other ways, particularly from loans and grants towards training and machinery expenditure in development areas, sanctioned by the Industrial Development Act 1966.[10]

Notwithstanding the ambiguities in state policy towards this sector, the interaction of leading producers and governments in the 1960s and early 1970s resulted in the transformation of the British textile industry's formerly fragmented structure. Nevertheless, industrial restructuring based on the promotion of capital concentration and vertical integration by partial state initiatives proved to be an inadequate solution to the long-term competitiveness of the textile industry. In just over two decades since the idea of 'infant-industry' protection was promoted, quota restrictions have spread from cotton to all fibres, while tariff protection against low-priced imports has grown. More importantly, it appears that the dominant strategy pursued by the large textile groups in the industry, emphasizing mass-market production based on a high degree of vertical and horizontal integration was not sufficient to make the industry competitive on a long-term basis against imports of standard, basic yarns and fabrics. The reasons why such a strategy was not a permanent success can be illustrated by analysing recent events at one company, TextileCo, and in particular its spinning division.

The Contemporary Restructuring of TextileCo

Textile production is unusually concentrated in Britain and is now dominated by three large corporate groupings. TextileCo is one of these groupings and its behaviour in the textile sector forms the background to the following discussion.

TextileCo is one of the largest British companies and, with world-wide sales of £2.3 bn and 65,000 employees, of whom 46,000 work in Britain and 19,000 in some 25 countries (Company Report 1986–7), is a major multinational by world standards. It embraces the entire textile and clothing production process from the manufacture of synthetic fibre, spinning, weaving, knitting, to garment assembly. In addition, it has extensive paint, plastic, and cellophane packaging interests, owns engineering subsidiaries and engages in wholesale and retailing. Nevertheless, behind this apparently buoyant corporate structure lies a very turbulent history. TextileCo discovered that a high degree of horizontal and vertical integration was extremely vulnerable in the volatile macroeconomic conditions of the 1970s and the early 1980s. With the oil crisis, stagflation, Third World and European imports, and competition from other synthetic fibre companies,

TextileCo found its growth sectors, such as the painting division, weighed down by increasingly unprofitable textile operations which constituted 70 per cent of total assets. By the start of the 1980s, a particularly unfavourable set of macroeconomic conditions (high interest rates, an appreciating pound, volatile energy prices) singled out an already weakened textile industry as a major target.

In Britain, textile production and fibre consumption fell by one third and employment by 165,000 or 37.6 per cent, from 1978–82 (Cable and Baker, 1983:115). The cotton sector (spinning, weaving) suffered a massive contraction. In mid-1978, 74,000 people were employed in the cotton sector and related textile activities in Britain; by mid–1982, this number had fallen to 43,000 (*ibid.*: 118). The adverse economic situation had grave implications for TextileCo. During 1981–2, domestic orders for the company fell by 20 per cent and exports by 21 per cent. This culminated in a financial loss of £114.2 million and the loss of 20,000 jobs in 1981, or one in four of the company's UK workforce. Eventually, this workforce was halved from its 1979 level to around 46,000 (Company Report 1986–7).

During the early 1980s, business observers argued that TextileCo was attempting to restore its overall profitability by cutting back capacity in its textile sector, whilst expanding into other, less cyclical areas such as paints and speciality chemicals. It was also suggested that the company was reducing its dependence on a volatile British market by refocusing its productive activities. From an analysis of capital employed by manufacturing location in the years 1979 and 1986, however, it appears that the company has not diversified to developing countries but to developed economies, particularly North America and Europe for cheap raw materials and geographical proximity to markets.[11]

But in spite of the ruthless rationalization of its textile sector in the early 1980s, the evidence suggests that the company is now implementing a restructuring strategy based less on savage retrenchment than an emphasis on the implementation of new technology and sophisticated marketing and sales techniques. This is made manifest by a £120 million programme of investment in the latest textile technology scheduled for 1986–9, which has coincided with a marked rise in the profitability of its textile sector. Indeed, during 1984–5, the profitability of the company's Textile Group rose by more than the total increase in the profitability of the whole corporation and in 1985–6, fibres, spinning and clothing had higher returns on investment than chemical and industrial products (Company Report 1985–6).

It is important to understand the basis for the revitalization of profitability in the textile sector. The company's Textile Group, in restructuring its production, is reacting to the reconfiguration of the fabric and garment buying system. Previously, this had been based on vertically integrated textile corporations, like TextileCo, co-operating closely with retailers, like Marks and Spencers, in producing standard garments and basic fabrics for a mass market at low prices. In the volatile, often recessionary conditions of the 1970s and early 1980s, retailers have responded to intensified competition by a range of sophisticated sales and marketing techniques. One strategy adopted has been to target particular segments of the market with loosely structured garment ranges which allows the individual consumer to construct the most favoured combination of stylish, colour co-ordinated clothes. This development sparked the rapid growth of the Hepworths chain, 'Next', whose co-ordinated fashionwear boosted sales from £20 million in 1983 to £70 million in 1984, while the Italian chain Benetton, adopted a similarly successful strategy aimed at a younger market. This approach has been imitated by other retail outlets, including Burton's, Marks and Spencer, British Home Stores and Littlewoods. Retailers serving the lower priced end of the clothing market, (C&A, Woolworths and the food chains, ASDA and Tesco), are also turning their attention towards increased variety and a more fashionable range of garments.

The new emphasis that retailers place on fashion and variety, however, inevitably affects manufacturers further down the textile and clothing production process. Yarn, fabric and garment manufacturers now require a high degree of productive flexibility to enable them to satisfy demand for a wider range of styles and to switch production between them in response to short-run trends in sales. This is inextricably linked to investment in the most modern textile and clothing technology to reduce unit costs of production and maximize flexibility to meet such varying customer requirements. Consequently, the market for clothing and its production is acquiring a dualistic structure. Capital-intensive sectors of the textile and to a growing extent, the clothing industry, which produce high quality, fashion fabrics and clothes now coexist with a labour-intensive sector of the garment industry employing women and ethnic minorities often in 'sweat-shop' conditions. Such manufacturers produce relatively poor quality, low-cost clothes to be sold by mail order firms, street traders, some large stores and a variety of independent retailers which are aimed at less affluent consumers (cf. Mitter, 1986; Chisholm *et al.*, 1986). The next section explores the response of the company's Spinning Division to these complex developments.

The Restructing of Production at the Workplace

The Company's Spinning Division has been heavily rationalized since 1979. During that year, it produced 1,800 tonnes of yarn a week and employed 11,000 people at 42 factories. By 1986, output was running at 1,100 tonnes with just 4,000 workers at 28 sites. Nevertheless, it still accounts for a hefty 55 per cent of the UK spinning industry and 26 per cent of all yarn consumed in the UK. In an effort to maintain and enhance its profitability, the Spinning Division has embarked on a programme of investment in new technology which has led to the complete re-equipment and modernization of one particular spinning plant.

Although it has been demonstrated that the state has had a profound impact on patterns of investment and the overall structure of the industry, the company's strategic decision to invest in new technology was not underpinned by government grants or available forms of industrial assistance. Since 1979, sectoral provision for the textile industry has been downgraded by the government in preference to private sector initiatives.[12] This paucity of state provision has been characterized in a recent analysis as a 'blanket refusal to give government assistance for industrial restructuring, although many other EEC countries have received large amounts of government aid and have maintained production and employment with more success than the UK' (Chisholm *et al.*, 1986:61). The most visible form of state intervention was the renewal of the Multi-Fibre Agreement in 1986 which marked the fourth phase of this type of import protection. This occurred after intensive and prolonged lobbying by both the textile and clothing industry associations and the trade unions through the TUC Textiles, Clothing and Footwear Industries Committee (cf. TUC:1985) and despite fears that the government would be reluctant to pursue a policy which ran counter to its ideological preference for the liberalization of international trade.

Hence, the reasons why a major programme of re-equipment has occurred at this particular mill will be analysed by reference to three crucial sets of relationships; intercapitalist competition and the relationship between workers which mediate the overall relationship between capital and labour.

InterCapitalist Competition
Both intensified global competition from developing countries and developed economies in the textile industry, and domestically based retail competition, have influenced the company's restructuring strategy. The company's Textile Group has responded to these developments by building a stronger position in its key markets by upgrading product quality and production technology. More specifically, the Spinning Division has embarked on a 5-year move to yarn production embodying a higher value-added component, enhanced style and design, and a full range of high quality, colour co-ordinated yarns for knitting and weaving customers with fast customer service at low cost. Standard carded yarns for cheap end uses have been discarded because of high import penetration, particularly from developing countries. Competition is now based on a more developed product range and not on a pure price basis, emphasizing proximity to the customer and quick service.

Intensified competition in the British retail market has, however, led directly to the re-equipment and modernization of one of the Division's spinning plants. Marks and Spencer, which takes 20 per cent of the Textile Group's sales, is constantly demanding better quality from their garment and fabric manufacturers. They now insist that their suppliers install more sophisticated technology. This machinery requires larger packages – or reels – of knot-free, high quality yarn, which has resulted in the recent investment decision by the company to spend £4.5 million to re-equip the plant with Schlafhorst Autocoro machines – the latest, fully automated open end spinning machines. These are to be gradually phased in over two years culminating in a 14-machine plant with new prespinning machinery, robotic package removal systems and advanced air conditioning equipment, and complete modernization of the building. This development will have profound consequences for labour. At a time of high unemployment labour productivity will more than double and since TextileCo does not foresee an increase in the factory's output of 70 tonnes of yarn a week, employment will fall by more than a half, from 250 to 100 employees. One positive benefit is that pay is likely to rise, although it is as yet unclear how productivity gains will be split between higher profits and higher wages. This epitomizes recent trends in British manufacturing industry of rapid growth of average earnings but continuing contraction of employment (64,000 jobs were shed in 1985 alone).

Senior management argue that the principal function of the technology was not to displace high-waged labour, as wages are low at the mill. Union officials estimate that workers at the Division earn basic hourly rates of £2.15 for a 37.5 hour week. With overtime, better paid workers gross £125 to £150 a week. Given the reconfiguration of the domestic retail and clothing market-place, and the intense competition from yarn manufacturers of other developed economies, management argue that they have little choice but to invest in the latest spinning technology. Any failure to introduce such technology would lead to erosion of market share and job losses. Hence, senior management saw the choice as between 'fewer jobs and a future and no future at all'.

Nevertheless, technology that more than doubles labour productivity will, in the long run, have serious consequences for employment levels. The extension of this type of technology to other parts of the Spinning Division will lead to redundant workers who cannot be absorbed by relocation or natural wastage. Such a strategy could prove inescapable if these current competitive pressures persist, and which, if ignored, would destroy the modest remnants of the Lancashire cotton textile industry.

Workforce Divisions
There have been no obstacles to the implementation of such a restructuring strategy posed by craft workers or strategically powerful bargaining groups within the mill. Technological change is not a new phenomenon in the textile industry and there are few work processes left which are intrinsically 'skilled'. The union concerned felt there was less 'skill' in-

volved in all work processes and this was inextricably linked to increasingly capital-intensive methods of production. Turner argued in 1962 (pp. 10–11) that 'few occupations in the cotton industry are intrinsically skilled in the sense that their adequate performance necessarily requires any long preliminary training. Most of the work is simple machine tending – feeding the machine with its material, removing its product, keeping it clean and free from obstruction.'

Even these are now defunct manual operations. The last bastion of an apprenticeship system at the mill was card attendants, a group which achieved skilled status by monopolization of union office and exploited technical change at the workplace to extend their control over machinery. This historic relic of skill will now finally be destroyed by new technology. Mill management felt spinning processes had been deskilled over many years by new technology and workers will get only minimal training for the new spinning machines. Both card attendants and draw frame operators (prespinning operations) will have their skill marginally reduced, although these jobs are only semiskilled. The only relatively skilled job that is created by the introduction of these machines is that of maintenance technician, and this position will be filled by a group of existing male supervisors. In short, new technology has not enhanced worker autonomy or increased task variety, but has further deskilled already routinized work processes.

There are also no strategically placed workers who could deflect or who would have any stake in resisting management's plan to re-equip the factory. Furthermore, certain workers will materially benefit from these developments and this is intrinsically related to the existence of a sexual division of labour at the mill. Pre-dating the introduction of the machinery, the work process was divided into the prespinning, spinning and winding processes. Prespinning processes – opening and cleaning, carding, draw and speed frame work – cumulatively clean and break up the dirty bales of raw cotton. The fibres are then straightened, separated and made to lie in a parallel direction and finally pulled and twisted to form a very soft, thick string which is ready for spinning. The mill had a section of ring spinning machines, which formed yarn by means of a spindle, and a small open end spinning section which produced yarn by centrifugal action. The winding process detects any faults in the yarn and finally, winds it onto packages ready for the customer. Generally, open end spinners are capable of higher productivity than ring spinners, although before recent technical advances, they have been economic for coarse rather than fine fabrics. But, many technical problems have been eradicated and Schlafhorst Autocoro machines – a third generation open end machine on the market since 1978 – have maximized the potential of open end spinning with full automation incorporating the winding process, and delivery of high quality yarn on large packages at speeds of up to 80,000 rpm.

Autocoro machines not only raise productivity levels (they are six times more productive than ring spinning machines) but they also eliminate certain work processes, notably the speed frame and winding operations and in this instance, will eliminate less productive technology like the ring spinning and the small open end spinning section. This has interesting implications for the sexual division of labour at the mill. Hitherto, the opening/cleaning and carding functions were a male preserve, as was the open end spinning section. Draw, speed frame work and ring spinning were performed equally by men and women, with winding an exclusively female preserve. All production workers were employed on a shift system, with the open end spinning section on a slightly different pattern, involving weekend work and slightly longer hours which resulted in higher pay rates.

In terms of the relative position of men and women, therefore, it is clear that male workers are less adversely affected by technical change. Firstly, there has always been a higher proportion of men at the mill because the open end section has worked a shift system which involves night and weekend work, considered by both management and unions to be unpopular with women workers. This shift pattern will be generalized to all workers in future. In fact, shift working, which gained ground in the 1960s, was emphasized by management as the principal reason for the transformation of the industry from a female to a male dominated workforce. Permanent night shifts or weekend work were held to be inconvenient for women because of their 'domestic responsibilities'. Futhermore, in relation to negotiations over the introduction of this technology, union officials had not considered negotiating shift hours which would explicitly suit the needs of working women with children.

More fundamentally, the people selected to be operatives on the new machinery will already be open end spinners who have the requisite 'skills', despite it being a semi/unskilled job. It is also the case that the opening/cleaning and card room section is also predominantly male, and will remain so with the introduction of Autocoro machinery. In fact, all the operations to be phased out are ones which predominantly affect women – speed frames, ring spinning and winding.

The single case of reskilling resulting from the re-equipment (maintenance technicians), will benefit male supervisors from whose ranks recruitment will come. The union estimated that the workforce at the mill will be 85 per cent male on higher pay, working in cleaner, modernized conditions. This is an example of the way in which male workers can and usually do capture the bargaining spoils of new technology because of the historical evolution of a sexual division of labour. Futhermore, the disproportionate displacement of women by technological change at the mill reflects complicated, long-term shifts in gender composition at the level of the industry.[13]

Hence, it has been argued that the heterogeneous composition of the workforce at the mill has enabled male workers to receive material rewards from the new technology in the form of higher pay and better working conditions. Men benefit from the re-equipment of the mill, both in terms of the change in work processes the new technology brings, but also because management and union officials seem unaware of the inherent discriminatory attitudes and practices that impregnate the negotiating process. This reinforces the conclusion of a Trade Union Research Unit (Moore and Levie, 1981) investigation into the impact of new technology in a range of companies. It appeared that older workers, Asian and female workers suffer most from technological change. The case studies indicated that new technology reinforces inequalities between men and women and sharpens divisions between unions.

Capital and Labour
The previous discussion has illustrated the ways in which intercapitalist competitive pressures and historically structured divisions between workers, have provided a positive context for the introduction of new technology at the mill. The prevailing relationship between capital and labour has also aided its introduction.

The union perspective is conditioned by the recent decimation of its membership and individual worker's attitudes, too, have been inevitably affected by high unemployment and the drastic rationalization of the Lancashire cotton spinning industry (between 1974 and 1985, the membership of the Amalgamated Textile Workers Union fell from 45,243 to 15,331 – ATWU, 1985). Hence, the union at an official level and workers at the plant have not opposed the introduction of new technology, although the union is pledged to fight any compulsory redundancies. In particular, the chance of higher pay and better conditions for the remaining workers and the possibility of redundancy money, which about half the displaced workers wish to take despite the offer of redeployment, has overcome any potential resistance by workers. This contrasts strongly with the usual impression of unions as blocking the introduction of new technology.

The union's benign response to technological change also reflects historical factors. The Amalgamated Textile Workers Union, established in 1974, was the result of a merger between five former separate unions, who had a predominance of membership in the cotton and allied textile spinning and weaving industries. Despite this amalgamation, the union kept a federal structure with two distinct levels of autonomy; the central organization (the Amalgamated Textile Workers Union) and the 10 districts which comprise the Amalgamated Textile Workers Union, each of which has its own set of rules and system of internal government with members' grievances and problems being processed initially at this level.[14] Turner (1962) in a study of the cotton unions, argued that despite some major disputes in the industry, the leaders of the cotton unions did not view industrial conflict as an instrument of broad social or political change, or strikes as anything more than an occasional necessity. Another factor was the cotton unions' overreliance on full-time officials. In comparison with engineering and dock workers, unofficial shopfloor based movements were rarely found in the cotton industry, and the level of workshop disputes was negligible. Turner argues (1962:28–9), moreover, that union leaders and employers colluded to limit the occasion for factory disputes by elaborating a general code of principles for their resolution. Hence, these historical antecedents – political conservatism, overreliance on full-time officials, weak traditions of shopfloor resistance and the peculiar federal system of district autonomy – have all shaped the character of the Amalgamated Textile Workers Union.

Moreover, the ATWU's response to technological change is not unique within the British union movement. British unions have shown little formal hostility to new technology and have generally co-operated with large-scale rationalization over the past seven years. Daniel (1986:264), in a recent survey concerning workplace industrial relations and technical change, concluded that 'trade unions did not operate as an obstacle to technical change' and that 'the general reaction was support for change, and often enthusiastic support', despite the fact that levels of negotiation and consultation with manual and non-manual workers and shop stewards were 'remarkably low'. Indeed, many unions have accepted the introduction of new work processes which have led to job losses rarely achieved through formal agreements and have not held rigidly to negotiated 'no redundancy' clauses (cf. Manwaring, 1981; Benson and Lloyd, 1983; Robins and Webster, 1982; Williams and Steward, 1985). According to Robins and Webster (1982:8) unions have accepted technology on management's terms 'conceding to the technological juggernaut they have "responsibly" accepted management's emphasis on profitability, efficiency and international competitiveness as the criteria underpinning technological innovation'.

For those workers who live in communities devastated by unemployment, however, and who work in a drastically rationalized industry which only just survived the recent recession, the option of resisting new technology is not a rational strategy. In the case of TextileCo, the workers are aware that their mill is an antiquated factory full of obsolete equipment, and that workers in surrounding mills would probably be more than willing to accept the new technology. In fact, contrary to Robins' and Webster's (1982) indictment of union acquiescence to the 'technological juggernaut', it is in this context understandable why workers do not oppose new technology when this response could jeopardize the competitiveness of the mill and existing jobs with few tangible gains.

Nevertheless, although at a micro level this particular group of workers was acting 'rationally' in accepting new technology in the context of the long-term decline and recent decimation of the industry, this is not to suggest that accommodation to the consequences of technological change is a 'rational' strategy for the larger collectivities of the union or textile workers generally. In the instance of the mill, not only did the union's strategy towards new technology perpetuate the social segmentation of the work-force, resulting in the disproportionate displacement of female workers, but it also illustrated an acceptance of the imperatives and exigencies of capitalist competition. Having accepted the argument that TextileCo needed to implement this type of technology to gain a competitive advan-tage in the textile market-place, its introduction then became a politically uncontested inevitability. The lack of any perceptible union strategy to deal with technological change and its long-term consequences in the textile and clothing industries not only reflects the present political weak-ness of the union movement generally, but is also related to the historical development of a consensus between employers, unions and the state regarding the underlying reason for the industry's decline and strategies for renewal. This consensus has traditionally coalesced around the adoption of protectionist measures to stem the rising levels of imports, especially of cheap clothing from the Third World, and thereby to protect jobs. How-ever, given the present restructuring of the industry, this appears an almost irrelevant strategy for the unions to pursue. Moreover, if the introduction of new technology is part of a more complex restructuring of the industry, and is being planned as a programme of investment at corporate level, then its negotiation at the level of the plant or establishment reinforces the isolation of both workers and union and seems to indicate that a coherent, long-term strategy needs to be formulated, not just by one individual textile union, but by all unions who represent workers in the textile and clothing industries.

Conclusion

The aim of this chapter has been to highlight and illustrate the complex nature of capitalist restructuring in the textile industry, both in terms of changing international patterns and the construction of corporate strat-egies at the level of the individual company. The analysis examined the strategies employed by one British manufacturer to restore the profitability of its textile sector by restructuring production towards upmarket, high quality products involving investment in new technology. As far as the company's Spinning Division is concerned, part of this strategy entailed the complete re-equipment of one of its mills. Elson's (1986) study of another British textile multinational – Tootal – also highlights the multiple strategies adopted by such firms to enhance profitability. In the case of Tootal, this ranges from investment in new technology in its UK clothing operations, commercial subcontracting in the Far East, and the buying up of existing productive assets operating at cheaper sites in the Far East and South East Asia.

In terms of a theoretical conceptualization of the restructuring process, it was argued that existing models, such as 'new international division of labour' and 'flexible specialization' are unable to deal adequately with the industrial changes and developments characteristic of today's textile industry. For example, 'new international division of labour' theories over-emphasise one pattern of restructuring – the geographical relocation of capital in search of cheaper labour – and therefore fail to appreciate fully the tendency of capitalism to expand through the technical reorganization of the production process. 'Flexible specialization' models accurately depict certain changes occurring in the textile industries of developed economies, notably the restructuring of production away from basic, standardized goods towards upmarket, high quality products for targetted segments of consumers. This does not mean, however, that such developments, which conveniently coincide with technological innovations, yield positive benefits for labour in terms of skill enhancement or increased control of the labour process. Moreover, this type of restructuring is intrinsically related to sophisticated product differentiation strategies adopted by the textile, clothing and retail distribution industries, in an attempt to restore and enhance profitability, and cannot be explained simply in terms of changes in consumer tastes.

Given the limitations of existing theories, the complex dynamics of capital restructuring in the textile industry were analysed in relation to several sets of relationships which are structuring contemporary developments and the particular strategies adopted by the company concerned. By a process of mutual interaction, they have generated a positive context for restructuring strategies based on the technical reorganization of production.

The historical background to contemporary developments in the industry and its highly concentrated corporate structure were analysed by reference to the partial and incomplete attempts by previous governments to restructure the industry and thereby stem its long term contraction and decline. Since 1979, however, the Conservative government has adopted a less interventionist role and restructuring has been the result of mainly private sector initiative, often at great social cost to textile workers who have been massively displaced from the industry. Hence, the strategy of re-equipment and modernization adopted by the company concerned and implemented at one particular spinning mill has not been underpinned by any form of state financial assistance. In short, three sets of relationships have interacted to encourage the company to undertake a technical reorganization of the production process.

The first of these relationships – intercapitalist competition – is changing at both an international and domestic level. Although studies have highlighted a substantial relocation of textile and clothing production to lower waged developing countries, it is arguable that this type of restructuring will become increasingly unattractive. An obvious trend towards upmarket restructuring in both the textile and clothing industries has meant that an increasing share of output consists of products which do not directly compete with the output of Third World countries. For developed economies, there is a need for proximity to centres of demand because of the dual requirements of flexibility and high quality production. Another major development has been an acceleration of the pace of technical change amongst developed countries with firms opting for the most automated, labour-saving technologies. This has provided individual firms in relatively high labour-cost locations with the possibility of resisting competition from developing countries by drastically reducing unit costs of production.

At the level of the individual company, TextileCo has adopted a similar upmarket strategy, which not only reflects the force of intercapitalist competitive pressures, but also stems from the changing nature of competition at retail level, which has induced retailers to upgrade their image and move away from low-priced, commodity garments. This has had a major impact on textile and clothing manufacturers. The new emphasis on fashionwear and varied, stylish, colour co-ordinated clothes demands enhanced flexibility to manufacture a wider range of styles and switch production between them in response to short-run trends in sales.

Contrary to 'flexible specialization' theorists, new technology, implemented as part of a strategy to enhance productive flexibility, has not had positive implications for worker autonomy or skill enhancement in the context of this case study. The new machinery further deskills already routinized work processes, apart from a single case of reskilling which is related to the creation of the new post of maintenance technicians. Hence, there are no strategically placed workers who could deflect or who would have any stake in resisting management's plan to re-equip the factory. Moreover, there is a distinct group of workers who will materially benefit from these developments. The existence of a sexual division of labour at the mill has ensured that male workers have been able to assert a prior claim to the new technology, and will receive the material benefits accruing from its introduction in the form of higher pay and better working conditions.

Intercapitalist competitive pressures and the nature of the divisions between workers have provided a positive context for the introduction of new technology at the mill and have mediated the overall relationship between capital and labour. It has also been affected by historical and economic factors which have conditioned the union's response to the introduction of new technology. The union has been affected by the recent decimation of its membership, particularly since 1979, but also by historical factors, particularly political conservatism, a uniquely federal and autonomous structure which would make co-ordinated, industry-wide action difficult and, lastly, weak traditions of shopfloor militancy. Workers' attitudes, too, have been affected by high unemployment and the drastic rationalization of the Lancashire textile industry and the company's Spinning Division in particular.

Generally, neither workers nor union officials have opposed the introduction of new technology as such. In any case, the mill is at present full of antiquated machinery and appears an obvious candidate for closure. For these workers, resigned acquiescence to technological change is a 'rational' response, given the long-term decline of the cotton textile industry. Nevertheless, the lack of any coherent strategy at union level to the problems posed by technological change and its associated job losses will become increasingly problematic given the 'technological' basis of much of the restructuring occurring in the industry.[15] This lack of strategic thinking, moreover, is intrinsically related to a disabling consensus which operates between the unions and employers and, since 1979, more equivocally the government, regarding the necessity of protectionist measures to stem the industry's decline and job losses by restrictions on low cost imports. Hence, it is argued that these factors have encouraged, not deterred, a restructuring of production based on the adoption of the most sophisticated and advanced set of production technologies.

Notes

1. The case study is based on interviews with the senior management of the spinning division, plant management and officials of the Amalgamated Textile Workers Union. These interviews were undertaken for an MA Dissertation during the summer of 1986. The interview data are supplemented by information from the trade and financial press; the company's Annual Reports; trade union (i.e. Amalgamated Textile Workers Union) and General Federation of Trade Union publications.
2. Multinational companies in the electronics industry are also pursuing complex, restructuring strategies (cf. Kyoung Cho, 1985).
3. According to Soete (1984: 167–8), 66,690 jobs were displaced as a result of textile imports from the EEC, compared to 18,189 due to Third World imports. More significantly, the impact of falling demand and productivity increases generated 146,320 job losses out of a total of 200,600 job losses in textiles.
4. The public inquiries were (1) the Report of the Cotton Textile Mission to the United States of America, March–April 1944 (London: HMSO, 1944). (2) Board of Trade Working Party Report: Cotton (London: HMSO, 1946).
5. The Textile Council – Cotton and Allied Textiles: A Report on Present Performance and Future Prospects (Manchester, 1969) Vol. 1 para 576.
6. According to Miles (1976:191), government expectations of the extent of re-equipment based on the industry's own calculations of probable costs were £80–£95 million, whereas total eligible re-equipment expenditure equalled £53.5 million.
7. As a supplement to the 1959 Act, voluntary quota agreements on imports of cotton textiles from Hong Kong, India and Pakistan were reached with the British cotton industry.
8. Knight (1974:104) states 'in January 1964 arrangements were made for a

meeting at which Sir Richard Powell, then permanent Secretary at the Board of Trade, urged Courtaulds to prepare a memorandum putting forward the case for a new approach.'

9. The Long Term Arrangement for International Trade in Cotton Textiles was negotiated in 1962 and was designed to give the textile industries of advanced capitalist countries a 5-year interval to adjust to intensified international competition. This has now developed into a permanent system of import protection in the form of the 'Multi-Fibre Agreement', which covers all the major textile fibres. It is designed to limit the volume of exports of textiles and clothing into designated EEC 'importer' countries from developing countries and the Eastern European bloc.

10. Such provisions encouraged Courtaulds to build weaving mills and a spinning unit in development areas. According to Knight (1974:175,182) 'without the financial incentives these large projects may not have gone ahead and certainly would not have been located in these areas.' It is estimated that about 25 per cent of Courtaulds' investment in the five years to March 1970 came from government grants.

11. This pattern of investment is not untypical. Grahl (1983), in an analysis of German direct investment, argues that its overall pattern is not primarily determined by the prospect of cheap labour in developing countries. He points to the importance of an international division of labour within Europe with a few Northern European countries tending to occupy the most advantageous positions, while routine unskilled tasks are dispersed to a Southern European periphery whose economic dependence is thereby increased.

12. There are only a few forms of sectoral assistance directed specifically towards the textile industry. The Textiles and Clothing Education Equipment Scheme, launched in June 1986, provides discretionary assistance to educational institutions for the purchase of advanced textile and clothing equipment. The European Communities' BRITE programme provides support for basic research in established industries such as textiles and includes a specific budget for research on the handling of flexible materials. A more promising measure was the 'CLOFT' scheme announced in March 1984 by the Department of Trade and Industry. £20 million was to be channelled into grant aid towards the purchase of technologically advanced equipment over a 4-year period. However, the European Commission announced in February 1985 that the Scheme was to be rejected because it was judged incompatible with EEC Competition Policy. Textile and clothing firms may also apply for assistance available under Sections 7 and 8 of the Industrial Development Act 1982, which includes Regional Aid and National Selective Assistance, but recent analysis of this type of provision concluded that 'these schemes cannot be said to represent a targetted approach to meeting the needs of the industry' and that 'the absolute levels of assistance under the Industry Acts have been progressively cut since the return of the present government in 1979' (Totterdill and Pearce, 1986:16).

13. Despite the rapid decline in the number of mule spindles traditionally operated by men, and an increase in ring spinning which was female dominated, the proportion of men employed in spinning has been rising since the end of 1959. This can be attributed to the growth of shift working, involving night work, intended to increase the efficient utilization of machinery at a time of declining

competitiveness. Women were prohibited from night work unless exemptions from the legislation were obtained, and this led to the gradual substitution of white female labour by Asian males. Generalized reductions in employment were also occurring because of the use of more automated technology, and mill closures. Such factors were clearly responsible for the decrease in female representation on union committees (Lewenhak, 1977:264). In particular, the union responsible for negotiations at the mill (ATWU) had an executive and district secretaries which were overwhelmingly male dominated.

Both Lewenhak (1977:263–4) and Walby (1986:232) argue that state policies had an impact on the sexual division of labour in the industry. In particular, the Evershed Commission (1945), set up by the government to investigate wages and work organization in cotton spinning, legitimated the substitution of male labour for female in certain occupations.

14. Due to the rapid and continuing decline in union membership, the Amalgamated Textile Workers Union merged with the General, Municipal and Boilermakers Union in 1986 to become its Textile Division. It is clear that such a merger will produce profound changes in the structure and character of the old cotton workers' union.

15. Even the highly labour-intensive sectors of the garment industry are now faced with a new phase of automation in the guise of 'flexible manufacturing systems' in which an integrated, computer automated system controls the complete production of garments. Nevertheless, it is unlikely that small firms in the clothing industry will be able to afford this type of expensive capital investment.

References

Amalgamated Textile Workers Union. 1985. *Eleventh and Final Annual Report.*

Arrighi, G. 1978. 'Towards a Theory of Capitalist Crisis', *New Left Review.* No. III, September-October, 3–24.

Benson, I. and J. Lloyd, 1983. *New Technology and Industrial Change.* London: Kegan Page.

Cable, V. 1977. 'British Protectionism and LDC Imports', *ODI Review.* No. 2, 29–48.

Cable, V. and B. Baker. 1983. *World Textile Trade and Production Trends.* Special Report No. 152. London: Economist Intelligence Unit.

Chisholm, N., N. Kabeer, S. Mitter and S. Howard 1986. *Linked By The Same Thread. The Multi-Fibre Arrangement and the Labour Movement.* London: Tower Hamlets International Solidarity.

Clairmonte, F. and J. Cavanagh. 1981. *The World in their Web: Dynamics of Textile Multinationals.* London: 2ed.

Corden, W.M. and G. Fels. (eds) 1976. *Public Assistance to Industry: Protection and Subsidies in Britain and Germany.* London: Macmillan.

Daniel, W.W. 1987. *Workplace Industrial Relations and Technical Change. Based on the DE/ESRC/PSI/ACAS Surveys.* London: Pinter/PSI.

Elbaum, B. W. Lazonick, F. Wilkinson and J. Zeitlin. 1979. 'The Labour Process, Market Structure and Marxist Theory', *Cambridge Journal of Economics.* 3, 3, 227–30.

Elson, D. 1986. 'The New International Division of Labour in the Textile and Garment Industry: How Far Does the 'Babbage Principle' Explain It?', *International Journal of Sociology and Social Policy*. 6, 2, 45–54.

Evershed Commission. 1945. *Report of the Evershed Commission on Wages and Conditions of Employment in the Spinning Industry, London: HMSO*.

Fine, B. and L. Harris. 1979. *Rereading Capital*. London: Macmillan.

Frobel, F. J. Heinrichs and O. Kreye. 1980. *The New International Division of Labour*. Cambridge: Cambridge University Press.

Gibson L.J. 1984. 'More Automation as Pace Quickens in Spinning Race', *Textile Month*. January, 15–36.

Grahl J. 1983. 'Restructuring in West European Industry', *Capital and Class*. No. 19, Spring, 119–41.

Guy, K. (ed). 1984. *Technological Trends and Employment 1. Basic Consumer Goods*. Aldershot: SPRU/Gower.

Jenkins, R. 1984. 'Divisions over the International Division of Labour', *Capital and Class*, No. 22, Spring, 28–57.

Knight, A. 1974. *Private Enterprise and Public Intervention: The Courtaulds Experience*. London: Allen & Unwin.

Kyoung Cho, S. 1985. 'The Labor Process and Capital Mobility: The Limits of the New International Division of Labor', *Politics and Society*. 14, 2, 185–222.

Landsberg, M. 1979. 'Export Led Industrialisation in the Third World: Manufacturing Imperialism', *Review of Radical Political Economics*. 11, 4, Winter, 50–63.

Lazonick, W. 1979. 'Industrial Relations and Technical Change: the case of the self acting mule', *Cambridge Journal of Economics*. 3, 3, September, 231–62.

Lewenhak, S. 1977. *Women and Trade Unions: An Outline History of Women in the British Trade Union Movement*. London; Benn.

Manwaring, A. 1981. 'The Trade Union Response to New Technology', *Industrial Relations Journal*. 12, 4, July-August, 7–26.

Miles, C. 1968. *Lancashire Textiles: A Case Study of Industrial Change*. CUP NIESR Occasional Paper XXIII.

Miles, C. 1976. 'Protection of the British Textile Industry', *Public Assistance to Industry Protection and Subsidies in Britain and Germany*. Ed. W.M. Corden and G. Fels, 184–214.

Mitter, S. 1986. 'Industrial Restructuring and Manufacturing Homework:immigrant women in the UK clothing industry', *Capital and Class*. 27, Winter, 37–80.

Monopolies Commission. 1968. *Man-Made Cellulosic Fibres*. London: HMSO.

Moore, R. and H. Levie. 1981. *The Impact of New Technology on Trade Union Organisation*. Report on research sponsored by DGV, Commission of the European Communities Study 1138–81. Oxford: Ruskin College.

Murray, R. 1985. 'Benetton Britain: The New Economic Order', *Marxism Today*. 29, 11, November, 28–32.

OECD. 1983. *Textile and Clothing Industries:Structural Problems and Policies in OECD Countries*. Paris: OECD.

Piore, M. and C. Sabel. 1984. *The Second Industrial Divide: Possibilities for Prosperity*. New York: Basic Books.

Robins, K. and F. Webster. 1982. 'New Technology: a survey of trade union response in Britain', *Industrial Relations Journal*. 13, 1, Spring, 7–26.

Shepherd, G. 1981. *Textile Industry Adjustment in Developed Countries*. Trade

Policy Research Centre. Thames Essays No. 30.

Shepherd, G. 1983. 'Textiles: New Ways of Surviving in an Old Industry', *Europe's Industries: Public and Private Strategies for Change*. Ed. G. Shepherd, F. Duchene and C. Saunders, 26–51.

Shepherd, G., F. Duchene and C. Saunders (eds). 1983. *Europe's Industries: Public and Private Strategies for Change*. London: Frances Pinter.

Soete, L. 1984. 'Textiles', *Technological Trends and Employment 1. Basic Consumer Goods*. Ed. K. Guy Aldershot: SPRU/Gower, 125–73.

Streeck, W. 1986. *Industrial Relations and Industrial Change in the Motor Industry. An International View*. University of Warwick: Industrial Relations Research Unit.

Textile Horizons. 1986. 'Spinning and Weaving Labour Costs', May.

Totterdill, P. and S. Pearce. 1986. *Prospects for Local Authority Intervention in the Textiles and Clothing Industry*. Centre for Local Economic Strategies Research Report No. 4.

Toyne, B., J.S. Arpan, D.A. Ricks, T.A. Shimp and A. Barnett. 1984. *The Global Textile Industry. World Industry Studies: 2*. London: Allen & Unwin.

Trade Union Congress. 1985. *A Fair Balance in Textile and Clothing Trade*. TUC Statement on MFA 4, May.

Turner, H.A. 1962. *Trade Union Growth, Structure and Policy; A Comparative Study of the Cotton Unions*. London: Allen & Urwin.

Walby, S. 1986. *Patriarchy At Work*. Cambridge: Polity.

Wilkinson, B. 1983. *The Shopfloor Politics of New Technology*. London: Heinemann.

Williams, R. and Steward, F. 1985. 'Technology Agreements in Great Britain: a survey 1977–83'. *Industrial Relations Journal*. 16, 3, Autumn, 58–73.

Zeitlin, J. 1985. *A Strategy for Local Government Intervention in the London Clothing Industry*. Report for the Industry and Employment Branch of the GLC. London: GLC.

Automotive Components

This chapter continues the theme of corporate responses to intensified international competition and changing product markets and, as in the case of textiles, considers the bases of one firm's relative success in restructuring.

The trajectory of the British automotive component industry's development has been heavily influenced by government policy in relation to motor vehicle production. For the case study company, the contraction of the British car industry and inward investment by foreign multinationals have compelled both retrenchment in the UK and the search for new markets. Intense competition between motor vehicle manufaturers, moreover, has forced the firm as a parts supplier into both product and process innovation.

A major multinational, the firm has adopted 'just-in-time' (JIT) production at its UK facilities, a system of manufacturing closely associated with the high productivity performance of the Japanese automotive industry. JIT combines new forms of labour control within production with new relations of vertical dependence between supplier and buyer companies. Work reorganization, changes in factory layout and in production control procedures enable the firm to diversity its product range whilst retaining the unit cost profile associated with mass-production. Often regarded as a low-cost route to productivity gains, the study describes the changes in labour management on which these productivity advances are based, and illustrates that JIT can prepare the ground for automation.

In place of the extreme division of labour associated with mass-production, job descriptions are drawn to encompass a variety of tasks, workers are regrouped into production teams, and new merit payment schemes replace individualized piece work. The aim is to maintain work effort by securing the internalization of control and the substitution of group pressures for direct supervision. Union job regulation is undermined in this process, as older solidarities based on demarcation disappear, reducing controls over the pace and distribution of work.

Because it is an 'ultra taut' production system, however, and one that increases the interdependence of firms, JIT is susceptible to the withdrawal of co-operation, either internally or in the firm's supplier or customer companies.

5

Industrial Restructuring and Labour Relations in the Automotive Components Industry: 'Just-in-Time' or 'Just-too-Late'?

Peter J. Turnbull

The Automotive Components Industry

Despite a massive rationalization of productive capacity in recent years, the automotive industry is still one of the most important manufacturing sectors of the British economy. It is the largest single source of manufacturing output, it plays a leading role in the development and utilization of new technology (such as computers, computer-aided design and manufacture (CAD/CAM), and robotics), and over one million workers are dependent on industry for their employment.[1] Furthermore, the mass production of largely standardized consumer products such as motor vehicles is widely regarded as the foundation of post-war economic prosperity; commercial success was based on mass production and competitive pricing, achieved through a particular form of work organization that facilitated 'economies of scale' and low unit cost production. The assembly lines that characterize the automotive industry were taken to be the acme of capitalist organization of the labour process, encapsulating the principles of 'scientific management' through a rigid form of technical control associated with specialized machinery and machine pacing, and a well-ordered form of social control associated with an extreme subdivision and fragmentation of work tasks, along with the spatial separation of mental and manual labour. In recent years, however, traditional forms of work organization and established patterns of management–labour relations have been challenged by economic crisis and the erosion of profitability throughout the industry. The competitive success of Japanese corporations in particular has provided the impetus for a new 'organizational paradigm' within the industry.

By the late 1970s, it was widely recognized that the production methods of the UK automotive component manufacturers (and vehicle manufacturers) were relatively inefficient. Many firms were operating old and unsuitable plant, which had an adverse effect on the ability of firms to organize the optimum layout of production facilities, and the decline of the indigenous motor vehicle industry had, for many firms, reduced batch sizes below the optimum level for existing scale and methods of production. This problem was exacerbated by the component sourcing policies of multinational car manufacturers based in Britain (namely Ford, General Motors/Vauxhall and Peugeot/Talbot) who increasingly sourced components on a European rather than British basis, and increasingly supplied the UK market through the assembly of 'knock-down kits' imported into the UK from overseas component and vehicle plants. As a result, the market for original equipment supplied to the vehicle manufacturers declined by a third over the past decade, while in the aftermarket scale economies were further eroded by the rapid growth of imports which created a plethora of vehicle marques and models and therefore greatly extended the range of replacement equipment.

When profits finally collapsed in the early 1980s, the response of the UK component manufacturers, almost without exception, was to cut excess capacity and manpower. However, 'crisis management' obscured a number of more important changes taking place within the industry, which can be explained by reference to the two basic market segments – the original equipment (OE) market and the replacement equipment (RE) market. The major component companies, the independent proprietary parts manufacturers who primarily supply original equipment and (usually) replacement equipment to the vehicle assemblers, have partially accommodated the rapid decline of UK vehicle manufacture and the consequent loss of volume for their products by exporting components from the UK and by extending or establishing production facilities overseas to supply foreign markets directly. They have also extended their presence in the aftermarket, but neither strategy has been sufficient to offset their ultimate dependence on the UK motor industry and their collapse of profitability in the early 1980s. Many smaller component firms also produce original equipment, usually to the specifications and designs of the vehicle assemblers, but many of these firms are essentially subcontractors and often do not even own their own tooling (ownership is retained by the vehicle manufacturers). Their contractual dependence on the motor manufacturers, combined with their lack of resources, has precluded the option of either an export or an overseas expansion programme, and consequently many have declined in tandem with UK vehicle production and new contractual arrangements that favour proprietary component suppliers.

The most profitable segment of the component market is the replacement market and a number of firms sell exclusively to the aftermarket, producing smaller volumes but at an enhanced profit margin. These firms have faced a very different set of problems created largely by the influx of imported vehicles and the consequent extension of their product range. Their problems have been further accentuated by their exclusion, until recently, from the franchized sector of the replacement market controlled by the vehicle assemblers and importers, and by the creation of additional competition from the vehicle manufacturers (who have developed an 'all makes' programme to supply parts for foreign as well as their own vehicles) and proprietary parts manufacturers (who have sought to increase their presence in the more profitable aftermarket by setting up their own distribution and service networks). Analysis of all four groups of component manufacturers – proprietary, OE subcontractors, vehicle manufacturers and RE manufacturers – and their activities in both market segments is therefore pertinent to an understanding of industrial restructuring and labour relations in the industry as a whole. Indeed, the four groups not only compete in the same markets, produce the same products and supply the same customers, they are currently involved in a 'zero-sum game' of 'winners and losers' where the success of one group or company is increasingly achieved at the expense of another. Present trends suggest that the large properietary component firms will be the survivors of restructuring within the industry, leading to both a greater concentration of capital and greater 'vertical dependence' between vehicle assemblers, proprietary parts manufactures, and subcontractors.

This chapter focuses primarily on the proprietary parts manufacturers, and in particular the recent experience of one of Britain's largest independent component manufacturers. Not only do the proprietary component companies account for the bulk of UK component production (the largest 20 independent component makers, excluding the motor manufacturers, account for around three quarters of total UK production, HCTIC, 27 November 7:9, Ref.91-i), their influence is likely to increase if present trends such as single-supplier contracts, shared research and development costs, and overseas expansion continue. Furthermore, the regeneration of the industry through a process of both destruction and reconstruction of capital equipment, the development of new products and processes, and above all the adoption of new management techniques, appears to favour the larger (multinational) component manufacturers. These companies are able to offer both the volume and the technical back-up required to secure contracts from the major vehicle assemblers, and they also have the resources to improve the quality and delivery dependability of their supplies.

Product quality and delivery reliability had been identified as the major areas of weakness of UK-produced components by the Price Commission on car parts in 1979. Some 10 per cent of purchases made by the four major motor manufacturers were delivered late or were deficient in some respect and suppliers often 'guessed' what would actually be required rather than produce and deliver to schedule, usually because of frequent changes imposed on them by the vehicle assemblers (Price Commission, 1979:57–9; see also Pendlebury, 1984:26). The Price Commission claimed that industrial disputes, and in particular unofficial stoppages, were the major cause of unreliability of delivery, as well as having an adverse effect on product quality (*ibid.*, 1979:61–2). Narrowing the 'industrial disruption differential' between the UK and its overseas competitors was therefore seen as the major problem for the automotive industry (*ibid.*:102), but the Commission failed to recognize the extent to which these problems were in fact a manifestation of the system of (mass) production, rather than simple 'labour relations problems'. Militancy has obviously been subdued since 1979–80 by high levels of unemployment, the loss of union membership, plant closures and mass redundancies, but there has been a more fundamental attack on improving product quality and delivery dependability through the adoption of JIT production and inventory control systems. Component companies are now expected to deliver (defect free) parts 'just-in-time' to meet the vehicle assembler's production schedule (see HCTIC, 1987), which enables cost savings through a massive reduction of stock, a speed up of the circuit of capital, and wide-ranging improvements to labour productivity. In addition, the organization of production under JIT is designed specifically to suppress industrial action and to circumscribe, if not eradicate, the autonomous role of trade union organization at the workplace.

The 'Automotive Component Company' (ACC) which forms the focus of this chapter now has well-developed JIT programme throughout its manufacturing plants. For ACC JIT is seen as vital to the viability and regeneration of its UK manufacturing plants. But the adoption of JIT is also regarded as fundamental to the regeneration of the automotive industry as a whole, both for component manufacturers and vehicle assemblers. JIT is therefore analysed in detail, both in reference to the experience of ACC and the experience of the components sector as a whole. Particular attention is given to the implications of JIT for union organization and representation at both the industry and plant/company level, as well as the effects and broader implications JIT has for work organization, the job tasks that workers are expected to perform, and the system by which they are remunerated. To understand the central role ascribed to JIT, however, it is necessary to locate its introduction firmly within the context of economic crisis in the UK automotive industry and the challenge from overseas competitors.

Market Competition, Economic Crisis and Industrial Concentration – the Scylla and Charybdis of Industrial Restructuring

Market Competition

Since the late 1970s, the world automotive industry has been suffering from a massive overhang of excess capacity. In 1982 the American auto industry was running at 50 per cent below capacity (*Investors Chronicle*, 22 January 1982:133), while current excess capacity in the European industry is in the region of 24 per cent (or 2.4 million units) for passenger cars and a massive 45 per cent for commercial vehicles (CVs) (Rhys, 1985). Despite this level of excess capacity and the decline of UK vehicle production, the productive capacity of the British components industry in the early 1980s was still geared up to supply a motor industry which, at its peak in 1972, produced 2.33 million units (1.92 million cars and over 400,000 commercial vehicles). Domestic car production fell below the 1 million mark in 1980 for the first time in over 20 years and, by 1984, the volume of UK components supplied to the UK car manufacturers was only a third of its 1972 level. Between 1972–9 this collapse was due almost entirely to the collapse of UK car production (illustrated in table 5.1). But after 1979 imported components accounted for at least half the volume loss (Jones, 1985), a process accelerated by the Conservative goverment's deflationary monetarist economic policies (high interest rates and a strong pound sterling on foreign exchange markets post-1979 reduced competitiveness of the UK automotive industry by approximately 50 per cent between 1979–82). The sudden collapse of production and the loss of international competitiveness left the components industry operating with between 20–30 per cent spare capacity in the early 1980s.

Volume loss attributed to declining domestic production was accelerated and exacerbated by the growth of imports into the UK during the 1970s (for example, the combined share of the three major Japanese motor manufacturers – Toyota, Nissan and Honda – increased from less than 0.5 per cent of the UK market in 1970 to over 10 per cent in the 1980s), and by the sourcing and production policies of UK-based multinational car manufacturers. These corporations increased both their overseas sourcing of components for cars assembled in the UK and the number of tied imports shipped to the UK from their European plants. Thus, although imported vehicles accounted for 57 per cent of new registrations in 1984 (SMMT, 1985:61), if tied imports of finished cars and components are included, then UK share of 'British' cars sold in the UK was just 34 per cent. This compares with 90 per cent in 1969, 69 per cent in 1974, and 39 per cent in 1979 (Jones, 1985). Austin Rover (AR) still sources around 85–90 per cent of its bought-in components from UK manufacturers, and Jaguar sources over 80 per cent (compared with over 90 per cent in 1980), but because of their European sourcing policy and the import of kits from

TABLE 5.1
World Car and Commercial Vehicle Production (000s)

YEAR	UK Cars	CVs	Europe[a] Cars	CVs	USA Cars	CVs	Japan Cars	CVs
1960	1353	458	3588	481	6675	1194	165	316
1970	1641	458	8156	830	6550	1733	3179	2110
1975	1268	381	7499	822	6717	2272	4568	2373
1979	1070	408	9600	1017	8434	3046	6176	3460
1981	855	230	8302	1035	6253	1701	6974	4205
1983	1045[b]	245	9376	979	6781	2424	7152	3960
1985	1048	266	9418	1034	8185	3468	7647	4624

Source: Society of Motor Manufacturers and Traders (1986)
Note: [a] includes France, West Germany, Italy and Spain
[b] 53 weeks

Europe to be assembled in the UK, both Ford and Peugeot/Talbot have a UK component content in their vehicles of less than 50 per cent, while the content of General Motors/ Vauxhall is less than 30 per cent (Jones, 1985; Birmingham EPG, 1987).[2]

The scale of excess capacity that afflicts the automotive industry has fuelled competition within the industry to a degree not experienced since the industry's major shake-out of the 1920s but, despite massive job losses and the rationalization and consolidation of productive capacity, the process of industrial restructuring and plant reorganization is far from complete. The problem of excess capacity will persist for some time even if demand picks up: imports into the European (and American) market are increasing, especially from Japan; there has been a marked fall in the level of direct exports from Europe (by 1.5 million units in the last 10 years) (Rhys, 1985); and *extra* capacity will continue to appear in Europe with further improvements to productivity, the consolidation of capacity by many successful 'specialist' producers such as Volvo and BMW and the likely entry into this market by the Japanese manufacturers, and the introduction of new assembly plants such as the Nissan plant in Northeast England. Retrenchment, rationalization and restructuring will therefore continue to be key features of the automotive industry in years to come but, as table 5.1 clearly illustrates, the British motor vehicle industry is *already* retrenched and rationalized. Indeed, among the major capitalist vehicle manufacturers the UK has suffered the greatest decline in both vehicle production and component manufacture.[3]

The decline of the British automotive industry is quite staggering. In the 1960s, for example, passenger car production was an important growth sector for the British economy, and the industry enjoyed a healthy export trade which constituted over 14 per cent of total UK exports (SMMT,

TABLE 5.2
British Automotive Industry Trade Balance (£m)

Year	Cars and CVs	Parts and Accessories[a]	Other Products[b]	Total Surplus (Deficit)
1970	409	338	181	928
1975	308	756	440	1054
1978	(548)	867	454	773
1980	(1014)	1053	554	593
1981	(971)	906	534	469
1982	(1843)	481	389	(973)
1983	(2827)	127	303	(2397)
1984	(2935)	231	392	(2312)
1985	(3205)	84	363	(2758)

Source: HM Customs & Excise, published by the SMMT (1986)
Notes: [a] includes rubber tyres and tubes, axles, shock absorbers, bodies, bulbs, brake and clutch linings, chassis without engines, electrical equipment, fan belts, engines, wheels, fans and blowers, crankshafts, plain shaft bearings, seats, radios and tape players
[b] includes industrial trucks, tractors, dumpers, trailers and caravans, and agricultural tractors

1985). By the 1980s the industry contributed less than 7 per cent of total UK exports, and the automotive industry is now showing a record deficit on its trade balance, as table 5.2 indicates.

Up until 1982, the trade surplus from the sale of parts and accessories produced by the automotive components sector was always sufficient to offset the (deteriorating) trade balance for cars and CVs, which showed a deficit for the first time in 1977. But after 1982 that situation was reversed quite dramatically, reflecting both the accelerating decline of the vehicle manufacturers (and its 'knock-on' effect on the components sector), and the growing problems of the components sector itself. The long-run decline of the domestic motor manufacturers, in particular British Leyland (now Austin Rover), led to serious problems for component manufacturers in respect of achieving optimum efficient scale in the face of reduced volume. This led to a consequent lack of investment and automation as volume was frequently insufficient to merit automation, and the gradual loss of the technical resources necessary to compete in international markets (see Price Commission, 1979). In addition, falling volumes have led to a notable absence of automotive product engineering in the UK in recent years (*Engineer*, 1 January 1981:34 and *Financial Times*, 16 August 1985), leaving the domestic components industry with a baseload of ageing components and at a competitive disadvantage in international markets. A number of components are no longer manufactured in the UK, and many at the 'high tech' end of the component range such as petrol fuel injection

TABLE 5.3
Imports of Automotive Parts and Accessories to the UK (1980 and 1985)

(£000s)	1980	1985	% Increase 1980–1985
EEC	501,979	1,255,998	150
West Germany	218,412	714,318	227
Japan	10,002	125,063	1150

Source: HM Customs and Excise, Statistics on Overseas Trade (1980 and 1985)

systems and spark ignition system are increasingly sourced overseas (see *Guardian*, 28 February 1987 and HCTIC, 1987).[4] Britain's trade (im)balance with the EEC in particular has grown commensurately with this decline, and in 1985 the UK had a trade deficit on automotive parts and accessories of £92m with France and £322m with West Germany. Our EEC 'partners', in particular the West Germans, have supplied the bulk of the increase of imported auto components (see table 5.3), although Japanese manufacturers have achieved the most spectacular import penetration over the last five years.

In the short term, though, the collapse of Britain's trade surplus in 1981–2 was caused by sudden cuts to capacity in response to falling demand and deflationary economic policies. Component manufacture is a highly volume sensitive industry (see Price Commission, 1979:56; *Financial Times*, 26 May 1983; and ICC, 1984), especially original equipment manufacture. Reductions in output by the major vehicle manufacturers have an immediate knock-on effect on the components sector, both in terms of output and employment.[5] In the first six months of 1980 volume collapsed by almost a third in the components industry, and total sales of original equipment were down by 40 per cent on 1979. This precipitated a collapse of profit margins throughout the components industry, regardless of firm size or product specialization.

Economic Crisis and the Collapse of Profitability
Between 1979–82, only two of the largest 66 component companies experienced any growth of sales, while 34 of the largest companies experienced a drop in sales of over 10 per cent (ICC, 1979–83). ACC, for example, was forced to cut production by over 20 per cent in 1980, and like the rest of the industry recorded a negative return on capital employed (ROCE) in 1981 (see figure 5.1).

In 1981, the largest 100 component companies, who account for approximately 80 per cent of the market, recorded an overall loss on over £2 billion of sales, and by 1982 almost half of these companies were loss-making. Even those manufacturers who had diversified out of automotive component manufacture and out of the UK were unable to stem heavy

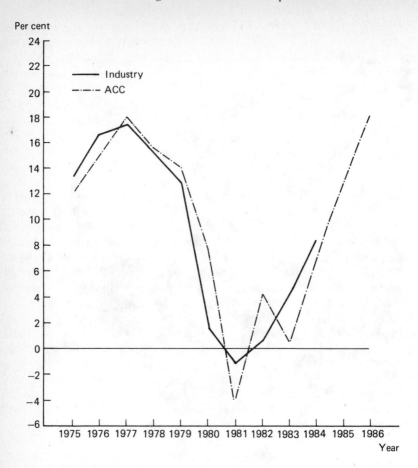

FIGURE 5.1
Return on Capital Employed, UK Automotive Components Industry

Source: ICC Business Ratios, Motor Components and Accessory Manufacturers
Notes: Return on capital employed (ROCE), defined as profit before tax divided by capital
employed, is the most widely accepted measure of capital profitability, indicating
how investors' money has been used to generate profits; data on the components
industry are based on the largest 100 companies

losses. ACC, for example, had reduced its dependence on automotive
components quite considerably. Automotive sales fell from around 80 per
cent to just over 70 per cent of total sales between 1978–82, although UK
production and UK exports still accounted for two-thirds of the total. In
addition, ACC had pursued a vigorous export policy, with exports as a
percentage of total UK company sales increasing from around 23 per cent
between 1975–8 to well over 30 per cent between 1981–6. But the company

still failed to realize a positive return on either its UK operations or its automotive business. The UK automotive operations earned a small profit before tax in 1986 for the first time since 1980, having lost well over £100 million between 1981–5. Only the company's overseas operations realized a positive profit margin in recent years.

The large proprietary component manufacturers, whose results dominate figure 5.1, have been trapped by their reliance on OE contracts with the motor manufacturers (Austin Rover, for example, still accounts for over 10 per cent of ACC's total sales).[6] Of course, a number of major UK component companies partially offset the decline of UK vehicle production by exporting from the UK and by establishing production facilities overseas to strengthen their presence in international markets (in 1985–6 there were 36 UK-owned component manufacturers with a turnover in excess of £10m p.a., and of these 23 had overseas operations, mainly in Western Europe and North America). But the vast majority of UK component suppliers do not have any significant export business, and throughout the industry there has been a general failure to rationalize production and invest in new products and processes by the late 1970s. At ACC, for example, employment was still around 90 per cent of its 1970 level in 1978–9, and the company had continued to add more capacity to its productive base in the UK. The principal factor behind this widespread 'managerial inertia' within the industry was the continued profitability of component manufacturers (see figure 5.1. In the early 1970s, ROCE was generally in excess of 10 per cent and profit margins around 6–8 per cent.) Although costs were rising during the 1970s the components industry was 'singularly successful in passing on these costs' (ICC, 1977), and was therefore able to maintain an acceptable level of profitability. Companies that had begun the process of rationalization and new investment in the late 1970s were suddenly hit by rising interest rates on bank loans that greatly increased overheads from debt servicing post–1979 (by 1981 ACC's liquidity had deteriorated by over £180 million), and when their price competitiveness was eroded in the early 1980s and the motor manufacturers refused to accept further price increases, even those companies that had managed to sustain a positive sales growth faltered. Component manufacturers as a whole were no longer able to realize an acceptable return on capital, profits collapsed and jobs and capacity were decimated.

Although the profit crisis was precipitated by specific events which eroded the ability of component manufacturers to pass on costs and realize profits, the long-term malaise of the industry had deeper roots which can be located within the dynamics of capital accumulation. As with other industrial sectors and capitalist economies more generally, the driving force behind the process of capital accumulation is the procurement of profit rather than the satisfaction of demand. Firms therefore continued to accumulate capital as long as they could secure an adequate return on the capital they employed, which in the case of component manufacturers was achieved by passing on costs to customers (the Labour government sanctioned such price increases under its Prices and Incomes Policy during the mid- to late-1970s). In such circumstances there is no innate reason for the supply of goods (potential or actual) to balance with market demand, and insufficient demand to clear the market can coexist with systematic overcapacity and overproduction. For individual firms, however, the eventual crisis always manifests itself as a situation where demand is too low and costs too high, and as a result they will always seek to cut capacity and manpower. But the long-term restoration of profitability and growth requires more fundamental changes to the process of production and the institutional structure of the industry. It is precisely because economic crisis invariably involves a reorganization of the production of surplus value that industrial restructuring involves both the elimination of un-profitable firms and unprofitable capacity (thereby raising the average productivity and profitability of the capital that remains), and the reorgan-ization of individual firms through investment in new products, new tech-nology, and new forms of work organization (thereby raising the productivity and profitability of individual capitalists). The Scylla and Charybdis of industrial restructuring involves at one and the same time both the destruc-tion and reconstruction of plant and capital equipment.

Industrial Concentration
During the 1970s, the motor manufacturers had responded to problems of rising costs and disruptions to component supplies by pursuing a policy of 'multiple sourcing' for OE contracts. But 'Dutch auctions' were simply a short-term 'solution' to these problems, and in the long term served only to weaken further the competitiveness of the supplier network by further reducing volumes and discouraging technological innovations. The costs of additional stock-holding to guard against distruptions to supply, and the deleterious effects on optimum scale created by fragmenting component contracts, further eroded the competitiveness of UK automotive com-ponents. In mid-1980 BL found that it could buy 70 per cent of its bought-in components overseas at an average price of 20 per cent lower than from its UK suppliers. Consequently, the major UK motor manufacturers reversed their sourcing policy in the early 1980s.[7]

Moves to single-sourcing as means of reducing price and restoring volumes has been vigorously pursued by all the major vehicle manufac-turers, not least Austin Rover (AR). As Harold Musgrove, then Chairman of AR put it, 'we are willing to help our suppliers make the necessary changes and to give them enough time to do so. But some are simply never going to be up to it and we must cross them off the list' (*Financial Times*, 26 May 1983). By 1982 AR was already sourcing over 50 per cent of its individual components from single sources (*Financial Times*, 31 March 1982), and between 1982–6 AR reduced its number of 'preferred suppliers' from 1200 to just 700 (HCTIC, 18 February 1987:162, Ref.143-v). Component contracts now last for around 5–10 years, rather than 1–2 years, and component companies awarded 'preferred supplier status' can now expect 100 per cent of AR's business for the component(s) in question. For example, Dunlop is now AR's sole supplier of wheels (to the detriment of Rubery Owen, who were subsequently forced to close their Daralston plant with the loss of 1,000 jobs); TI Fulton is now AR's sole supplier of brake pipes (to the detriment of Armco); and Lucas Batteries is the sole supplier of around 500,000 batteries per annum to AR, despite a complaint to the Monopolies Commission registered by Chloride who formerly supplied half the contract. AR's policy appears to be paying off. By 1982, the price differential between UK and overseas component manufacturers had fallen to around 12–13 per cent (*Financial Times*, 31 March and the *Engineer*, 21 January 1983:7); many individual component have subsequently been subject to a total price freeze (Bessant *et al.*, 1984:61); between 1981–2 AR's material costs increased by only 2 per cent (*Financial Times*, 26 May 1983); and the quality of components supplied to AR has greatly improved (HCTIC, 18 February 1987:162, Ref.143-v).

Other UK motor manufacturers have reduced their suppliers by a similar magnitude. Land-Rover now has only 840 suppliers and intends to reduce the number still further, Ford UK has reduced the number of its (European) suppliers from around 2,500 in 1982 to just 900 in 1987, and Nissan intend to single-source all its components from fewer than 150 suppliers by 1990 (HCTIC, 1987). On the basis of volume alone, single-supplier contracts clearly favour the large proprietary component manufacturers, while medium and small-size companies are more likely to secure only OE contract for the lower overhead type of components (HCTIC, 18.2.1987:165, Ref.143-v). But it is not just volume that favours the larger component manufacturers. The motor manufacturers are also looking to share more research and development work with their suppliers (which, by definition, favours proprietary parts manufacturers), and many vehicle manufacturers are now doing less work on the design and material specifications of their bought-in components. Furthermore, design work will increasingly be linked through engineering facilities using CAD/CAM which many smaller manufacturers cannot afford, and the vehicle and engine manufacturers increasingly favour 'packaged' systems of components offered by the proprietary component companies (such as engine management systems, complete steering or suspension systems and braking systems), rather than individual components produced by specialist subcontractors or smaller independent manufacturers. The smaller (subcontractor) component companies that manufacture OE exclusively for the motor manufacturers are also disadvantaged by the rising costs of tooling and the growing reluctance of motor manufacturers to own such tooling leased to specialist component manufacturers (HCTIC, 21 January 1987:58, Ref.143-i). The smaller component manufacturers that do survive are therefore more likely to become subcontractors to the large, multinational proprietary component manufacturers, rather than direct suppliers to the vehicle manufacturers.

'Vertical dependence' is reinforced by another new development, namely the adoption of JIT production and inventory control systems. In Japan, where JIT systems are most fully developed and integrated between vehicle assemblers and component suppliers who make several component deliveries a day, there is a very high degree of industrial concentration and vertical dependence. This gives the Japanese automotive industry a 'pyramidal' structure. At the top are the major vehicle manufacturers, dominated by Toyota and Nissan who together account for well over 60 per cent of total car production (Price Commission, 1979:54). The vehicle manufacturers source around 75 per cent of their components externally (compared with around 48 per cent in the USA and between 50–60 per cent in Europe), virtually all from single suppliers. This in turn leads to a high level of concentration among component suppliers (the top 350 component manufacturers account for 90 per cent of final output, and 80 per cent of that output is sold directly to vehicle assemblers as either original or replacement equipment) (*ibid*.:60). The vehicle assemblers therefore retain a much tighter grip on the aftermarket with a market share of around 50 per cent, because smaller companies work for larger companies on a subcontract basis rather than manufacture parts for independent outlets in the aftermarket.[8] Below the major Japanese component manufacturers is a tier of subcontractors who number around 7,000 (Price Commission, 1979:60), but below that there is a tertiary layer of subcontractors who number tens of thousands and whose size diminishes right down to the individual, self-employed workers performing routine assembly tasks in their own homes. Thus, Toyota has around 170 primary suppliers, who in turn depend on over 4,000 secondary subcontractors, who in turn depend on a tertiary, outermost layer numbering many thousands (see *Financial Times*, 9 November 1983: Dodwell, 1986:32). Nissan has a similar network of suppliers, but its primary suppliers would generally be different from the firms supplying Toyota.

Vertical contractual dependence of this nature can have serious, if not dire consequences for both the structure and status of employment within the industry or economy more generally. For the major corporations vertical dependence gives rise to the organizational advantages of vertical intergration without the financial obligations, allowing them to offer their own employees more secure employment and superior employment conditions at the expense of workers employed by their subcontractors. The latter become 'shock absorbers' for demand fluctuations affecting the large corporations, and they can offer their employees only temporary employment status and inferior pay and conditions (see Turnbull, 1988). Thus it has been estimated that in Japan only 20 per cent of the total workforce is accorded permanent employment status (the so-called 'life-time employees'), 65–70 per cent are non-regular workers in smaller firms (usually the subcontractors to the large corporations), and around 3.5 million labourers are employed/unemployed on a daily basis (George and Levie, 1984:27; see also Hayes and Wheelwright, 1984:364). There are already clear signs of industrial concentration and greater productive flexibility achieved through subcontracting in the UK automotive components industry. But the implications of JIT are more pervasive than their effects on contractual dependence and the structure and status of employment. JIT fundamentally transforms the labour process and the pattern of industrial relations within the industry, creating a new technical and social structure of capital accumulation. The JIT system and its implications for labour are therefore elaborated in some detail.

Manufacturing Just-in-Time

Why Just-in-Time?

Just-in-time (JIT) refers narrowly to a particular way of organizing the manufacturing production process to secure the maximum return on capital employed by increasing the rate of 'throughput' through the plant (i.e. output per unit of time). The system is capable of marrying the seemingly incompatible objectives of high quality, low unit cost, manufacturing flexibility and delivery dependability, thereby distorting the 'competitive equilibrium' in many international markets. The competitive implications of JIT are so pervasive because it enables companies to use the cost savings associated with the system either to make the same product range at lower cost and/or higher quality, or make a greater variety of products (within limits) at a similar or even lower cost (see Abegglen and Stalk, 1985:117). For ACC and many other UK automotive firms, the system appears to be tailor-made to their needs, given the state of product demand and structure of component manufacturing costs. The principal problem facing ACC by the mid- to late-1970s in respect of manufacturing capabilities and competitive performance, which was perhaps typical of the components industry as a whole, was how to control manufacturing costs in the face of a proliferating product range and falling volumes. As the company's director of manufacturing operations succintly put it, 'during the 1950s ACC was making massive volumes in a few products, perhaps three or four. Now we are doing exactly the opposite; we are talking four hundred or so. But we are trying to cope with the new requirement from an old base.'

Increasing the product range is clearly an attractive *marketing* strategy for the firm, enabling the company to capture a wider market share or specialized market niches while simultaneously spreading financial risks. But such a strategy can upset the balance between marketing and manufacturing, creating additional problems for the latter in terms of increasing the number of component parts produced; increasing inventory levels and material handling expenditures; complicating process flows and generating quality control problems; reducing batch sizes and shortening production runs, thereby eleminating a principal source of economies of scale; committing greater overheads to indirect areas of production such as scheduling and expediting (progress chasing); and ultimately *increasing* unit costs of production. By the late 1970s, ACC was experiencing problems associated with over-complicated production scheduling, poor quality control (associated with degenerate payment-by-result systems in its major automotive plants), excessive lead times and a poor reputation for reliability of supply, and a cost structure that was incompatible with the changing demands of the automotive market. Unlike the company's Japanese competitors in particular, ACC was unable to meet the demands of the market and match its production schedules to the pattern of final demand. Instead, it was increasingly embroiled in the day-to-day problems created by the manufacturing process and delivery schedules, dictated largely by the need to produce large, homogeneous batches to cover operating costs.

The adoption of JIT production and inventory control systems is therefore regarded as the vehicle for restoring the compatibility between the demands of the market and constraints of the manufacturing process. Put simply, JIT is a system that aims to eliminate all elements of 'waste' in the manufacturing process, where waste is defined as anything (or anyone) that adds cost, but not value to the finished product. For ACC JIT is also seen as a means of establishing the viability of individual plants and products, as well as the profitability of its UK manufacturing operations in general. More importantly, however, JIT is currently being hailed as a panacea for British manufacturing industry as a whole (*Engineering Computers*, 1986:55–60), and it has been claimed that JIT is 'the most important productivity-enhancing innovation since Frederick W. Taylor's scientific management at the turn of the century' (Schonberger, 1983:13; see also Monden, 1981:36). But while the productive *potential* of JIT may be relatively uncontentious, the extent to which it is predicated on a specific form of industrial structure (vertical contractual dependence), and an associated form of labour relations is often overlooked (see Turnbull, 1988). In addition, a number of economic variables outside the control of individual firms may serve to nullify its potential for securing long-term profitability and growth. These aspects of JIT are evaluated in the subsequent analysis, but first the 'nuts and bolts' of the system are elaborated.

Just-In-Time Production and Inventory Control
With JIT, the exact quantity and uniform quality of (defect-free) raw materials, parts and subassemblies are produced and delivered *just-in-time* for the next stage of the production process (this principle is extended backwards to the company's suppliers and subcontractors who are expected to make several deliveries of materials and components during each working day). In other words, the time between materials intake and

goods dispatch is kept to minimum through arrangements that eliminate the need for raw materials inventories, 'buffer stocks' of work-in-progress between work stations or manufacturing processes, and finished goods inventories. The cost savings from such an approach can be quite considerable. For example, it has been estimated that in a typical Western manufacturing plant, materials and parts are worked on for only 5 per cent of the time they spend in the factory – up to 95 per cent of 'in-process-time' is spent moving the product between operations and queueing (see Ballance and Sinclair, 1983:148; *Engineering Computers*, 1986:55) – and not surprisingly 30 per cent of production costs in many Western plants go on warehousing, inventory carrying and monitoring (*Business Week*, 14 May 1985). Inventory reduction clearly produces significant cost savings and time economies in the circuit of capital, but to regard inventory control *per se* as the principal objective of JIT is to underestimate its competitive potential.

JIT production aims to minimize *all* costs surrounding the production process, not just those associated with high inventory levels. All unnecessary or 'wasteful' elements of production are eliminated, such as material queues which consume both time and space; stocks of any kind, since these not only take up valuable storage space and necessitate unnecessary handling time but also 'cover up' inefficiencies in the production process (for example, making the source of defective parts production more difficult to locate); production in excess of that immediately required or scheduled which, by definition, is not needed and is therefore wasteful; waiting time of machine operators; set-up time; and the production of defective parts or products. Numerous changes to the organization of production are therefore required to achieve these objectives.

The primary objective, at least initially, is the reduction of machine set-up and changeover times. Rather than amortize set-up costs over long production runs, firms can now produce smaller quantities of any particular product on an economically viable basis by minimizing changeover and set-up times/costs, enabling them to switch production more frequently, increase the variety of products, and reduce inventory levels and the idle time of the workforce. Firms employing JIT therefore invest in multi purpose machinery and tooling with quick changeover dies, calibrated machine tools that eliminate trial and error adjustment, and special jigs that facilitate the achievement of 'single set-up'(i.e. single digit number of minutes) or 'one touch set-up' (i.e. zero set-up time or only load and unload time). They also include machine set-ups in the job descriptions of direct production workers, rather than rely on a separate (craft) grade of tool setters.

Although JIT is often presented as a low – risk, 'low cost' option, to the extent of not requiring any major new investment in capital equipment (see Rice and Yoshikawa, 1982:1; Saipe and Schonberger, 1984:60; and Cook, 1984), investment in new tooling, CNC machines and electronic quality control systems have gone hand-in-hand with the introduction of JIT at ACC. In particular, because JIT is operated with the dual concept of 'total quality control', firms may find themselves involved in expensive investment programmes to ensure they 'get it right first time'. While it is difficult, therefore, to disentangle the productivity-enhancing effects of new tooling and new technology from the effects of JIT, it is apparent that the move towards just-in-time inventory control and delivery schedules reinforces the market position of large OE manufacturers who have the resources to invest in the technical supports necessary for an effective JIT system.

However, reduced set-up times should, theoretically at least, enable small and medium-sized firms to compete with the large proprietary component companies since it reduces the optimum batch size/production run. For example, by offsetting the economies of long production runs, firms can produce/assemble the same end products every day (Western manufacturers typically assemble products in batches, assembling a particular product for a set period, usually several days, before changing over to another product). 'Mixed assembly' under a JIT system enables production to be planned in such a way that the firm is more responsive to product market demand, and it enables production planning to achieve a more uniform *flow* of production within the plant. This then creates a more uniform demand for parts/subassemblies on all preceding processes. The production schedule for the whole plant can be determined by final assembly by 'homogenizing' the final assembly sequence and then minimizing any fluctuations in the final assembly stage, known as 'level scheduling' or 'smoothing production' (see, for example, Monden, 1981:40). The final assssembly schedule is therefore fixed or frozen for several weeks, and the demand for parts/subassemblies is allowed to 'ripple back upstream' through the production process. That is, the production rate of any one process determines the depletion rate and thus the rate of replenishment of components, parts or sub assemblies from the previous process (this also applies to components or raw materials sourced externally). The demand for parts is therefore 'pulled through' the production process, typically controlled by production order cards called 'kanban'.[9] In this way each stage produces, just-in-time, only that amount of parts necessary to complete the next stage of the production process.

Despite the potential that JIT offers for smaller component manufacturers, it is the large OE suppliers who have so far adopted JIT systems (see Tailby and Turnbull, 1987; Turnbull, 1988), often in response to contractual requirements imposed by the major vehicle and engine manufacturers. JIT is especially suited to 'continuous flow manufacture' or 'repetitive manufacturing' processes which characterize the large component companies because even when lot sizes reach the optimal level of one unit, as at Toyota, the idea is to produce one piece after another *continuously*.[10] Thus, despite the variability of products leaving the production line every day, the principal benefits of JIT appear to accrue primarily to fabrication and assembly processes, especially those that already employ repetitive manufacturing sequences (see Crosby, 1984:27; Saipe, 1984:41; Saipe and Schonberger, 1984:61; and Finch, 1986:31). This reinforces the tendency of OE component manufacturers to adopt JIT, but even for these manufacturers the major challenge of adopting JIT has been to transform job-lot or batch production processes into repetitive manufacturing operations resembling a continuous flow plant (see Monden, 1981; Spurgeon, 1983; Saipe and Schonberger, 1984; Schonberger and Schniederjans, 1984; Celley *et al.*, 1986). By reducing set-up times, improving quality control, eliminating or drastically cutting inventory of all kinds, and most importantly reconfiguring operations in a series to resemble a flow line, component manufacturers can create the rhythm of assembly-line pacing in a far wider variety of manufacturing settings.

Implementing Just-In-Time Production

The Technical Reorganization of Production
The most visible strategy of companies operating JIT systems is the reorganization of the factory layout. Machines are located in close proximity to each other to (physically) prevent the accumulation of inventory between work stations, enabling the firm to economize on storage and material handling costs. Machinery is also located in a 'line' (or U-shaped 'cell') in the order that machines are needed to complete consecutive stages of the manufacturing process such that they operate like an assembly line. This facilitates further reductions in set-up times, a compression of production lead times, and the instigation of new working practices such as teamworking, individual responsibility for quality and the maintenance of machinery, and multi-machine manning. This organization of the production process stands in stark contrast to Western manufacturers who typically organize their production process and factory layout according to *process* or *functional* groupings of machinery, where work passes from one process to the next interspersed by (prolonged) periods in work-in-progress inventories between each work station. With process groupings of equipment, the pace of each process is effectively decoupled from both subsequent and preceding operations by work-in-progress inventories. With JIT, and the regrouping of machinery into 'group technology', the pace of work is effectively coupled with all other processes because the firm can select families of parts or products that follow approximately the same production routing and regroup their machinery into 'product lines' or 'manufacturing cells' determined by these routings. This increases the rate of throughput by generating a more continuous *rate* of demand and a more consistent *flow* of production along each product line. Not only is production speeded up quite considerably, but the flow of production can be controlled and monitored more precisely.

The reorganization of the plant layout into several product lines or 'manufacturing cells' is now common throughout ACC's automotive plants in both the UK and overseas. A more detailed description of just one manufacturing plant will therefore serve to illustrate the basic principles adopted.[11] The plant in question manufactures engine equipment, and its experience was typical of many other plants within the company. By the early 1980s the plant was suffering from chronic excess capacity, caused by a sharp decline in its traditional market (agricultural and commercial vehicles), after a prolonged period of expansion during the 1970s based heavily on a strong export performance (the plant was exporting around 70 per cent of its finished products). Throughout its phase of expansion the plant was basically manufacturing one 'family' or type of component fitted to the slow(er) revving diesel engines manufactured for agricultural and commercial vehicles. Competitive success was based on high volume production of this standardized product, priced at a competitive level and mass produced at very high precision tolerances (machining tolerances are only 1 1/2 microns, or 0.00006", which is just one fiftieth the diameter of a human hair). But the decline of this market in the late 1970s, coupled with a rapid growth of new markets (namely diesel engined passenger cars) combined to reconstitute the company's entire product range by the early 1980s.

The traditional engine component manufactured for agricultural vehicles and CVs currently accounts for only 40 per cent of total production. The plant now manufactures six different component 'families', each requiring different manufacturing technology and process techniques, very often produced at much lower volumes than the traditional equipment. In total, the plant now manufactures around 300 different variants of its engine component. a proliferation commensurate with the range of vehicles now fitted with such equipment. Furthermore, legal regulation of fuel emmissions and fuel economy is now a major constraint on both the design and manufacture of its engine equipment for cars and light CVs, requiring even greater precision during machining operations and demanding more rigorous quality and test procedures.

The reorientation of the product market created innumerable manufacturing problems. When the company was manufacturing one basic engine component the manufacturing philosophy was quite simple – 'you just push five tons of raw material in at one end and three months later machined components would appear at the other end' (Plant General Manager) – but with a proliferation of products it became increasingly difficult to co-ordinate and monitor production flows, to locate the source of quality problems, and to meet customers' delivery schedules. These problems arose directly from the traditional 'process' form of work organization, which was typical not only of other ACC plants, but also other engineering plants in the automotive components industry. Machines would be grouped together according to the function they performed, work passed from one specialist machining centre to the next, and each group of machines would be 'owned' and supervised by a foreman. There was nothing inherently inconsistent or problematic with this layout when the company was manufacturing only a limited number of product variants, but with greater product variety the manufacturing system itself was proving untenable. In particular, there was insufficient information and communication between the interface of these process groupings, hindering the flow of production and making quality problems difficult to locate. These problems were strongly associated with the company's (degenerate) payment system and the structure of supervision – defective parts production concerned the workforce only to the extent that it prohibited the individual attainment of average piecework earnings, and with supervisors responsible for only a single process, it was often difficult to locate both the source and the cause of quality problems. It was decided, therefore, to assign foremen to two interconnected process groupings to improve communications and generate information on the interaction between machine groups. This principle was then extended to cover a discrete product family produced by a series of interconnected machines, effected by a total regrouping and re-layout of the factory floor. Work now flows along seven *product lines*, interrupted only by a heat treatment process to enable further machining, rather than following a zig-zag pattern between several *machine* groupings.

Product manufacture based on group technology uses the 'focused factory' concept, where the production system is designed specifically for a limited number and variety of production lines. This practice eliminates conflicts between the needs of different products, and has been likened to the creation of a series of autonomous, self-contained 'mini-factories within the factory'. The first product line established at the engine component plant began production on an experimental basis in the early months of 1984. The new system yielded a 25 per cent increase in productivity (output per employee), and total reorganization of the factory was effected over the next 12 months. Where volume was high enough it was possible to dedicate a block of capacity to a particular product family, as was the case with six of the seven product lines, but where volume was insufficient the solution was *mixed* production (line 7) where a larger number of products are manufactured every day in a mixed sequence (this line mainly manufactures RE equipment for non-current engine models). By mixing products in this way the same scheduled product mix can be produced almost every day, thereby creating a uniform flow of production similar to that of the other six lines.

As one might expect, the reorganization of the factory into group technology reduces the workers' ability to control the pace of work, since they can no longer build up 'banks' of work, or employ other devices that allow them rest periods at their own discretion, and with no work-in-progress stocks workers are tied together in a line and paced within very narrow limits set by the production process. In the initial reorganization phase the stock level was reduced from nine weeks supply to just ten days, throughput time was reduced from eight weeks to only seven days, the number of rejects at test was halved, machine utilization was increased from less than 70 per cent to around 90 per cent, and lead times were reduced from around three months to just ten days. As the company's Chief Engineer put it, 'nothing stands still anymore', and that includes labour. Richard Schonberger, a principal exponent of JIT, explains this aspect of JIT quite graphically – 'in the traditional batch production enviroment, the workers are like servers in a conventional restaurant; in the JIT environment they are like servers in a fast-food restaurant' (Saipe and Schonberger, 1984:64). Thus, greater control of production scheduling, and the drastic reduction of inventories, serves to 'ratchet up the level of stress at which the workforce is expected to perform' (Abernathy *et al.*, 1983:76; see also Sayer, 1986:66).

However, during the first phase of the plant reorganization – regrouping machinery from a functional to a group technology basis – employees were essentially performing similar tasks to those they had performed previously, albeit at a higher rate of throughput. Only the supervisory structure was radically altered, and various 'indirect' jobs such as quality inspectors, progress chasers and stores personnel were eliminated. But although plant reorganization brought a significant reduction in unit costs of production (by 11 per cent in 1984 alone), the plant was still operating with excess capacity (if the plant had been able to operate at full capacity unit costs could have been reduced by a further 20 per cent). In addition, the plant is still less competitive than its principal West German and Japanese competitors, against whom the plant gauges its performance. As a result, further changes to the manufacturing process have been implemented, basically centred around a total rewriting of work tasks, performance levels, and job responsibilities. As the company themselves put it, these changes require not just a fundamental redesign of manufacturing systems but also 'far–reaching changes in traditional working practices, habits and culture' (Company Report, 1986). Understandably, these changes have far-reaching implications for trade organization and the future conduct of workplace industrial relations.

The Social Reorganization of Production

The changes to work organization discussed so far have focused mainly on the technical reorganization of the labour process. But as the central objective of JIT is the reduction of wasted or inefficiently employed labour and machine capacity, the system is associated with a distinctive set of working practices and distinctive social controls of the labour process. Although these practices appear, at first sight, to be the very antithesis of traditional forms of work organization under the mass production paradigm, closer scrutiny indicates that JIT should not be regarded as an alternative to Taylorism 'but rather a solution to its classic problem of the resistance of workers to placing their knowledge of production in the service of rationalization' (Dohse *et al.*, 1985:128).

The principle underlying 'conventional' mass production systems is that the volume production of standardized items is the most effective route to securing lowest unit costs. In accordance with this principle, maximum labour utilization is sought through the extreme horizontal subdivision of labour and fragmentation of work tasks. Jobs are task specific, parts are moved between specialist work stations, employees perform a narrowly defined operation repetitively, and employees are remunerated according to the task they perform or the job they occupy. Quality is *tested into* production by a separate grade of inspection personnel and, since the production of defective parts is assumed to be inevitable, large buffer stocks are maintained to ensure the continuity of production during equally inevitable periods of downtime. With institutionalized demarcations between separate work groups, reinforced by both management and unions, the system can be extremely inflexible, especially in respect of periods of downtime or changeovers to different products when production workers must wait for maintenance workers to repair equipment or wait for a skilled setter to reset machine tools. It is the wastage represented by such downtime, or idle time more generally, that the JIT system aims to eliminate.

Maximum labour utilization under JIT is achieved by extending the principle of 'no buffer stocks' to labour – there is 'no buffer personnel'. Consequently, job descriptions are drawn more widely through, for example, the introduction of multi-machine manning and the reallocation of various indirect tasks to direct production workers. The latter now perform indirect tasks *on line*, and are held individually responsible for routine maintenance tasks, minor breakdowns on their machines, and the quality of parts they produce. With no buffer stocks the production system cannot tolerate the production of defective parts, since everything is being produced in the exact quantity just in time for the next stage of production, and therefore quality has to be *built into*, if not *thought into* the system (see Hayes and Wheelwright, 1984:361–70). The complementary concept of 'total quality control' involves considerable cost and productivity savings, not only from the elimination of indirect grades of personnel and greater/ maximum utilization of direct production workers, but also from the fact that throughput increases (defective parts do not occur and less 'unproductive' time is spent reworking parts); the production process itself is simplified because defective parts no longer have to be rerouted through the plant; and less waste means lower inventory requirements. A 2 per cent reduction in the defect rate of a machining intensive plant such as ACC's engine component plant can produce a 10 per cent increase in productivity through precisely these mechanisms (see Hayes and Wheelwright, 1984:363). In fact, most of the initial productivity improvements reported by the plant were attributed to improved quality control and simplified production control derived from 'getting it right first time'.

The wider implications of JIT, with its emphasis on problem-solving, teamworking, multi-machine manning and broader job definitions, is that employees are expected to move between different activities as the *workload* dictates, rather than at the insistence of (on-cost) supervisory personnel. The emphasis on 'teamworking' developed by Japanese companies is frequently cited in this context as an explanation for the willingness of Japanese workers to respond to this requirement for flexibility. More realistically, with no buffer stocks to fall back on, no demarcations between jobs, and flexibility across a wider range of work tasks, workers are 'exposed to continual, controlled pressure' (Hayes and Wheelwright, 1984:360) and are left with little option other than to 'co-operate' and 'use their initiative' to keep production going (Schonberger, 1983:14). In this way the costs of production irregularities are effectively shifted on to the worker such that social pressure among the work group to maintain the continuity of production and to attain (rigid) output quotas becomes a functional aspect of production control (this 'internalization of Taylorism' is discussed in detail by Dohse *et al.*, 1985:129–33).

At ACC the most important changes brought about by the initial reorganization of the factory layout into product lines were related to quality control, teamworking and a new group payment system which replaced payment-by-results (PBR). All three factors were closely interrelated. With foremen now responsible for the product rather than the process it became possible to nail down accountability and responsibility for defective production to an individual worker rather than a process. Individuals were thus made responsible for the parts/products they produced, and workers along each product line were recognized as a 'team' for the purpose of wage payments. The new team system based wage payments to the group on the number of non-defective finished components produced by the team's product line, with output quotas set at 80 per cent machine efficiency for each product line. This gave the team a level of earnings comparable to average piecework earnings under the PBR system (although some workers had to be paid 'make-up' lieu rates that were gradually phased out because the changeover involved a reduction in their average weekly earnings). The team payment system is intended to give all workers equal earnings potential, but more importantly it allows greater individual assessment of employees according to their skills or other personal characteristics. It could facilitate the adoption of individual, discretionary merit payments rather than a straightforward 'rate-for-the-job' (as at Nissan's plant in Northeast England), and it will serve to highlight the importance of the quality of production as well as the quantity of output.

Improved quality control was seen as a prerequisite of any reorganization of the production process, as vehicle assemblers and engine manufacturers are now demanding that their suppliers guarantee the quality of their products. Austin Rover and Jaguar have made Statistical Process Control (SPC) a contractual requirement, and Ford now operates its own quality system (Q101) to assess its component suppliers. ACC operates SPC throughout its manufacturing plants, with quality responsibility assigned to the production department rather than a separate test or quality control department. With individual workers made directly responsible for what they produce and the operations they perform, it is now possible to monitor production standards continually, enabling *trends* in production to be determined by means of mini-computers connected to machines and 'auto-activation' facilities which stop the machine at the end of each cycle, or when parts begin to exceed tolerances. This enables machine operators to perform pre-emptive adjustments to machines to eliminate the production of defective parts and to minimize downtime and waiting time. At the engine component plant every component is electronically tested at the end of each product line, and every result is stored as a permanent record. The new test sequence is actually completed in half the time taken by the old procedure, human error has been virtually eliminated, and a comprehensive record is now available to determine both quality standards and the team payments for each product line. Unlike traditional quality control, therefore, which relied on 'rectification loops' to rework defective components discovered by the test department, SPC controls and monitors quality throughout the production process (see *Engineer*, 13 February 1986:27).

The reorganization of the plant layout, along with the introduction of new technology such as CNC machines, electrical discharge machinery (EDM), and sophisticated electronic quality control devices, has allowed management to increase the level of work effort and redefine customary levels of active co-operation. Machines now stop automatically at the end of each production cycle or when a defective part is produced, and consequently it is no longer necessary for workers to 'baby-sit' machines. Multi-machine manning, whereby workers operate several machines and assist other workers in the team who are 'overloaded', facilitates an intensification of work effort and a higher level of machine utilization. In addition, workers who are 'overloaded' frequently come under pressure from their team colleagues to intensify their work effort and keep up with the pace of production, rather than displace additional responsibilities on other workers.

The changeover to product lines, teamworking and SPC at the engine component plant redefined the work of one craft group (the Standard Room) and eliminated a number of indirect jobs such as stores personnel, progress chasers and materials handlers. A new shift working pattern was also introduced, with a changeover from day and night shifts (7.30 a.m. to 4.30 p.m. and 8.30 p.m. to 7.00 a.m.) to a double-day shift system with two-week rotations (that is, two weeks on 6 a.m. to 2 p.m. then two weeks on

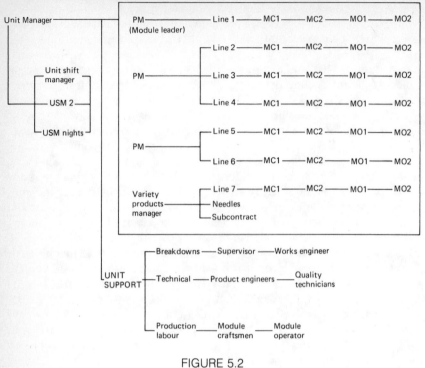

FIGURE 5.2
Work Organization

Notes: PM = Product manager
MC = Module craftsmen (two grades)
MO = Module operators (two grades)

2 p.m. to 10 p.m.) with a 'linking' nightshift (10 p.m. to 6 a.m.). But more significant changes to working practices have been subsequently effected by a reorganization of the work teams along each product line. Groups of (formerly distinct) craft workers have now been integrated into the product line teams, traditional areas of maintenance have been assigned to direct operatives (such as preventive and simple routine maintenance), job responsibilities have been extended and new working practices introduced (for example multi-machine manning), and the number of direct grades has been reduced from 15 (13 skilled and 2 semiskilled) to just four (two craft and two operator grades). Figure 5.2 illustrates the new structure of work organization within the engine component plant, where each product line is now denoted as a self-contained manufacturing 'module'.

ACC has always operated a policy of in-house sourcing of components, which meant that large areas of the factory and its workforce were given over to machining, pressing, casting and welding raw materials straight from the stockholders into the finished product. Each of these processes required its own technical and material back-up in the form of draughtsmen, engineers of various types, and special craft groups with the resources necessary to maintain existing processes and to develop new ones in line with new product developments. The introduction of JIT has squeezed out the specialist craft groups both from below via the deployment of 'manufacturing craftsmen' (who ideally will perform first-line maintenance in areas previously covered by electricians, pipe fitters, machine tool fitters and toolmakers, as well as performing their more routine duties such as first-line supervision, quality control and tool setting), and from above via the introduction of a new technical support unit drawn from the ranks of craft workers and technical staff employees. Although many craft workers have understandably been reluctant to join the module teams as manufacturing craftsmen, management at the engine component plant avoided any overt forms of conflict by recruiting to the modules on a voluntary basis. Because it took several months to reconfigure machinery and equipment into the distinct product lines, management were able to fill 'vacancies' on each consecutive line on a piecemeal basis. Consequently, the workers on line 7 (the last line to be reorganized) were the most reluctant to change over to the new system, but any collective resistance to the new working practices had been undermined over the previous months. After regrading all the jobs within the modules, management then 'readvertised' all production jobs and asked workers to 'reapply' for jobs in order that they could select those workers whom they believed to possess the requisite skills for the new job categories.[12]

Assigning greater responsibility to workers within the modules has necessitated an extensive assessment of training and retraining requirements. It is unclear, however, whether the new working practices such as multi-machine manning and self-inspection simply involve an increase in workers' *responsibilities*, or a genuine increase in the exercise of their *capabilities*. It is certainly the case that JIT systems involve greater employee responsibility and require greater levels of commitment, co-operation and self-discipline than do more traditional forms of work organization. But it does not follow that a workforce operating a JIT system must necessarily be more skilled. In other words, while there is definitely a greater emphasis on developing *behavioural* skills compatible with the productive objectives of the firm, there need not be any reskilling of the workforce or any greater degree of employee autonomy (see Dohse *et al.*, 1985; Sayer, 1986:67; Shaiken *et al.*, 1986; Turnbull, 1986, 1988). More important, therefore, is the extent to which these new working practices enable management to redefine the level of work effort and customary levels of active co-operation through teamworking and other practices that frequently by-pass existing channels of trade union representation.

Just-In-Time, Union Organization and the Conduct of Industrial Relations

Although JIT is associated with a specific set of working practices, these practices are by no means exclusive to companies operating JIT systems. Indeed, many British companies have recently introduced teamworking, quality-at-source, group incentive schemes and the like (IDS Study 360, 1986), and no doubt these firms will, in some instances, reduce costs significantly. But it is unlikely that their cost savings will be as great as companies operating a 'true' JIT system, especially where firms are simply taking advantage of favourable labour market conditions to erode traditional union restraints on output, manning and the way jobs are performed. Where this is the case, or where the focus is simply on shifting the wage–effort bargain by redefining job responsibilities, it is likely that new demarcations will appear (albeit at a higher level of flexibility), that unions will seek to tighten up controls on subcontracting, part-time or temporary working, and that they will insist on a much higher price for future changes. In fact, these trends are already evident at a number of companies that have recently negotiated 'flexibility agreements' (IDS, Study 360, 1986:22–4). JIT addresses the more general problem of improving the productive performance of the manufacturing system as a whole, rather than speeding up the work of individual workers or individual machines, and consequently changes to working practices, manning levels and the like produce more fundamental changes to the contours and conduct of workplace industrial relations.

If 'true' JIT systems are to be widely adopted in Britain, then the implications for traditional forms of workplace industrial relations are quite inauspicious. In Japan, for example, the JIT environment is one in which there are hardly any limits to managerial prerogatives, and the industrial relations system is one that limits the articulation of collective interests by employees. Peer group pressure has been substituted for more traditional forms of supervisory/managerial control on the shopfloor, while wage systems and career patterns based on meritocracy and strong competition between workers who lack any effective (collective) means of resistance ensures that the system functions with little difficulty or disruption (Dohse *et al.*, 1985:140–2). There are signs of similar changes in the sphere of industrial relations at ACC and other UK automotive companies.

JIT has served to undermine the conditions which gave rise to old-style demarcations associated with 'functional' technology and process layout. There is no longer any *technical* basis for separate craft groups, the majority of whom have been integrated into the manufacturing modules, and therefore no technical basis for separate negotiations and recognition for craft groups. The pay of those workers still classified as craftsmen (MC1 and MC2 grades) is no longer based on separate craft rates but is now tied to the performance of their team colleagues (both craft and non-craft workers) along their particular product line. Thus the basic unit of organization at the workplace has shifted from skilled craft groups and semi-skilled production workers to (managerially defined) product teams. Although the company has no short-term plans to press for single-union recognition, JIT and its modular format of work organization clearly lends itself to more 'rational' forms of worker representation or at least undermines the basis of multi-union representation. At 'greenfield' sites single-union representation is now the norm rather than the exception, but short of a 'Wapping solution' at most well-established 'brownfield' plants it seems unlikely that single-union representation will be negotiated. However, that does not preclude coercive tactics on the part of management when it comes to implementing Japanese-style working practices (see Turnbull, 1986, 1988; *Guardian*, 2 July 1987, 7 July 1987). If nothing else, therefore, JIT will involve new industrial relations practices and will no doubt necessitate a reorganization of shop steward representation, shifting the 'natural' constituency of steward organization from the craft or process group to the product team.

In some respects there is nothing remarkable, or even exceptional, about these changes to the structure of union recognition and representation. Taken in isolation, the rationalization of recognition, the elimination of craft groups with formerly separate negotiating rights, and a redefinition of the basic bargaining unit may appear fairly innocuous. But in conjunction with these changes, JIT involves more fundamental changes to the *social* organization of the labour process, and these changes are far more pernicious. With JIT the rationale for teamworking, flexibility between tasks, multi-machine manning, quality-at-source, self-inspection and group technology is not just to redefine the level of work effort and the customary level of active co-operation on the job, it is also intended to reconfigure the method and even the topics of interest articulation on the part of the workforce. With a process layout, management are frequently imbued with a 'myopic concern with speeding up *individual* machines and *individual* workers' (Sayer, 1986:48), which often serves only to create further production imbalances between individual work stations. With JIT systems, workers are tied together more closely through the coupling of machinery and tighter control of product flows, and thus it is intended that workers' attention should shift to those factors that determine (non-defective) output and therefore earnings of the product team.

In other words, teamworking and its associated work practices are intended to inculcate a sense of responsibility for the quality of the product, the speed and efficiency of production, and the achievement of production targets. In Japanese companies the attainment of these objectives is frequently associated with quality circles and other forms of direct management–worker communications. At the engine component and other ACC plants, the unions have strongly resisted quality circles, but management do not appear too perturbed by this resistance – as one manager put it, 'we don't really need quality circles – our quality system [SPC] takes care of quality control and involves workers in the operation and update of quality control procedures'. Furthermore, management also have a new line of direct communication with their workers at the shop-floor level via the product managers or module leaders, which by-passes the shop steward organization altogether. More important, however, is the fact that quality consciousness and responsibility for the achievement of production goals and standards is inculcated by generating peer group pressure among members of the module teams, that is, pressure to maintain required quality standards, to achieve output targets, to assist other workers in the day-to-day operation of the plant and to cover for absent colleagues (and therefore to be fully conversant with their work tasks), to rectify production problems 'on the spot', and to maintain a higher degree of co-operation and self-discipline within the work team. All of this ultimately determines group earnings. The ultimate objective of the work practices associated with JIT, therefore, is that they foster greater worker dependence on the firm (rather than the union) in respect of remuneration, promotion, employment status and job security.

Establishing greater employee dependence on the firm is perhaps the key to securing the full benefits of a JIT system. In Japan, for example, greater flexibility in the allocation of work tasks is often erroneously ascribed to teamworking, 'group culture' or the 'corporate ethos', when in reality individual assessment of employees and more individualized systems of remuneration are actually the key to securing full-scale job flexibility (since this ties 'core' workers ever more tightly to the firm) (Dohse *et al.*, 1985). Many companies in Britain currently regard the individual assessment of employees as a key barrier to achieving full flexibility (IDS Study 360, 1986:14), which is even more pertinent to companies such as ACC who are operating a JIT production system.

At ACC's engine component plant the process of establishing a more individualized system for pay and conditions of employment began in earnest with the reorganization of the plant into seven product lines when each line was manned on an individual, voluntary basis. Rationalization of the grading structure and the redefinition of job tasks reinforced this trend with workers allocated to the four new job grades according to managerially defined criteria of skills and capabilities. Consistent with this strategy is the establishment of a dual labour force. 'Core' employees can expect relatively more secure employment, but are expected to exercise greater versatility and adaptability in the tasks/jobs they perform. 'Peripheral' employees, in contrast, are often employed on only temporary or fixed-term contracts to give the firm greater flexibility in respect of product market fluctuations (for example in the peak production period prior to the August registrations of new vehicles). This principle applies both within plants and between AAC's plants and their external subcontractors.

At one ACC plant, for example, an automated casting process operated 24 hours a day employs only the engineers on the company staff, while the small manual workforce is provided by an external employment agency. It is very difficult for the trade unions to organize and recruit these small groups of manual workers, and the company estimate that a further 20 plants have a similar degree of total automation to enable similar employment conditions to be instigated. Further intra-plant flexibility is achieved by assigning all 'critical' manufacturing operations such as high precision machining to the core workforce, while all 'non-critical' operations are subcontracted to outside companies. JIT enables companies to identify these operations more easily, since it simplifies product flows and manufacturing operations, and it therefore adds a further dimension of flexibility and market responsiveness to the armoury of companies such as ACC who can effectively operate a JIT system.

White-collar workers have not been immune from this process either. JIT principles have been applied to information processing, paperwork associated with order processing, and the design function, and with no interstage work orders on the factory floor and only intermodule work-in-process monitoring required (rather than every work station), the number of staff associated with production control and data processing has been reduced considerably. In fact the redundancies of recent years have been most heavily concentrated among 'indirect'/non-production employees such as maintenance, stores personnel and staff employees, and management's most concerted action to reorganize bargaining procedures and rationalize recognition has been directed toward the company's white-collar unions. Unlike negotiations with the manual workers' unions, negotiations with white-collar employees are conducted at the national level and determine pay and conditions across all ACC plants. Management were intent on dismantling these arrangements in favour of plant level bargaining, or what they call a 'consortium' approach to collective bargaining. In 1986 the nationally negotiated pay settlement for white-collar employees was tied to the acceptance of flexible working conditions on a plant-by-plant basis, giving plant managers the authority to determine the new working arrangements and to decide whether or not the individual plant could afford the pay rise. Acceptance of the deal would have eliminated the nationally negotiated pay and grading structure which determines career progression within the company, to be replaced by a 'personalized' salary structure determined by management. Industrial action by the white-collar unions eventually produced a national pay increase, but the question of decentralizing the bargaining structure was not resolved. Management's determination to pursue the matter forced the unions to capitulate and agree to the break-up of the national pay bargaining forum, in conjunction with the harmonization of terms and conditions of employment across white-collar and manual employees, and the linkage of staff salaries to plant performance. However, the unions are continuing to resist the decentralization of bargaining over conditions of employment.

Within each product team, therefore, and among the company's white-collar employees, a clear hierarchy of pay and employment status is being developed, based on a more individualized system of assessment than hitherto, and determined predominantly by management. This is probably the most pernicious facet of the redefinition of the trade unions' traditional role within the workplace under a JIT system, undermining the tenets of altruism, collectivism and egalitarianism.

Conclusions: 'Just-In-Time' or 'Just-Too-Late'?

In 1979 the Price Commission on car parts identified several major weaknesses of the UK automotive components industry. Near the top of the list were product quality, delivery dependability and price competitiveness. Recent evidence from component companies and vehicle manufacturers suggests that the UK components industry has made significant progress in these areas (HCTIC, 1987), and that JIT has played a significant role in improving the competitive performance of UK components. But the major problem identified by the Price Commission, namely the effects of industrial disruption on competitive performance in the international as well as the domestic market (1979:13–14, 61–2, 102), has largely been forgotten. There are obvious reasons for this, not least those associated with the adoption of JIT and managements' subsequent attempts to break down traditional demarcations and collective bargaining arrangements that often precipitated short, unofficial stoppages. But although the level of industrial disruption may be much lower in the 1980s than it was during the 1970s, firms may now be more vulnerable to such disruption. Austin Rover, for example, now holds only four days stock for major bought-in components (Birmingham EPG, 1987), whereas UK vehicle manufacturers used to hold anywhere between one and four months stock (Price Commission, 1979:5). Industrial action at any one of the vehicle manufacturers' major OE suppliers can therefore bring production very quickly to a standstill, as the overtime ban and subsequent strike at Lucas Electrical in the Autumn of 1986 aptly illustrated (see *Guardian*, 30 October 1986).

Thus, while Japanese production systems and their associated personnel, employment and remuneration systems are widely regarded as generic rather than specific to culture or philosophy (Schonberger, 1982), it should be evident that the Japanese industrial relations system plays a pivotal role in the successful operation of JIT systems (see Dohse *et al.*, 1985; Turnbull, 1988). We must therefore question the extent to which such systems can be operated successfully in countries with a very different structure, organization and history of organizational strength and control of the immediate working environment, and to ensure the continued autonomy and independence of collective bargaining procedures.

This must involve recruitment of those workers who seem likely to be 'marginalized' by the adoption of JIT (namely workers on short-term contracts at the proprietary component companies and the secondary labour force employed by subcontractors to the major manufacturers); by the joint control of new demarcations among the workforce and the definition and assessment of workers' skills and other attributes; and by resistance to new forms of communication that explicitly exclude union representation. As Sayer (1986:69) points out, 'whether there is anything progressive for labour in the new working practices depends on the form they take and the form which labour *lets* them take' (emphasis added). But whereas union organization appeared to act as a brake on the reorganization of production in the 1970s (Price Commission, 1979:11), recent events highlight the extent to which trade unions have been unable to influence the nature of industrial restructuring at major component companies. While this is no doubt testament to adverse labour-market conditions in general, and continuing problems in the automotive components product market in particular, it must be emphasized that the JIT production system presents numerous inherent problems for trade union organization and interest representation.

JIT is aimed at securing employee involvement in the rationalization of production, or what might be termed an 'internalization of Taylorism' (Dohse *et al.*, 1985:128; Ichiyo, 1984:46; Sayer, 1986:68; Wood, 1986). Thus the emphasis on teamworking, employee motivation and high performance levels is not borne out by some predilection for the welfare or job satisfaction of the workforce, but with the need to break down the rigidity of existing production systems and to restore the firm to a situation of profitability and growth. JIT systems are therefore innately 'Janus faced' – while they may stress the importance of behavioural skills that promote co-operativeness, conscientiousness and self-discipline, JIT production systems are also highly oppressive, often securing high levels of productivity by overtly coercive means (Kamata, 1982; Dohse *et al.*, 1985; Sayer, 1986). Coercive elements have recently been seen to predominate in the managerial restructuring of production in the UK automotive industry (e.g. Willman and Winch, 1985; Grunberg, 1986; and Turnbull, 1986), despite, or possibly because of, low levels of worker and union militancy.

There is certainly very little scope for component manufacturers such as ACC to elicit higher levels of worker motivation, effort and co-operation on the basis of enhanced job security or enhanced levels of pay, the so-called 'carrots' of JIT that are frequently assigned an incommensurate role in explaining the flexibility and superior productivity performance of Japanese corporations. ACC finally made a small pre-tax profit on its UK automotive operations in 1986 after several years of sustained losses, but its programme of rationalization continues unabated. Plant closure, the rationalization of existing capacity into fewer manufacturing units, and the sale of ailing plants to overseas multinationals continue to be key features of the company's UK operations. At the same time ACC's European automotive operations recently increased sales by over 30 per cent and profits by over 40 per cent. The simple fact is that, despite improving the performance of its UK operations through the implementation of JIT, the domestic automotive market is unable to sustain viable scale economies. This is true for ACC and virtually every other major component company, regardless of product specialization (HCTIC, 1987). So although it is quite certain that there will remain successful, market-oriented companies serving the consumer sectors of replacement parts and accessories, 'what is not so certain is whether a motor components industry of the size and sufficient technological sophistication to retain its present stake of OE business, and expand it, can survive' (ICC, 1985:6).

Thus, despite improved flexibility and reductions to the size of efficient scale achieved through JIT, component manufacture remains a volume-sensitive and volume-dependent industry. The choice for component manufacturers is therefore quite stark:

> There are really only two possibilities for UK based suppliers. One is to develop strong international business links which could incorporate joint ventures and cooperative agreements with other suppliers or the establishment of their own multinational operations. The other is to shrink into a relatively peripheral second-line supplier of unimportant parts (TI Automotive, evidence to the HCTIC 18 March 1987:221).

The restructuring of the industry on management's terms alone, and on an uncoordinated basis, is unlikely to ensure the survival of a strong UK components industry, especially one that could support the entire product range of automotive components and in particular many of the new 'high tech' electronic components. The full benefits of a JIT system take anywhere between five and ten years to realize (Aggarwal, 1985:10), assuming the full co-operation of labour and discounting any significant labour relations problems. Industrial restructuring along these lines may well ensure the 'survival of the fittest', but 'economic Darwinism' is neither in the interests of labour nor the economy in general. Without wider political initiatives to support the automotive industry the process of perennial decline will continue unchecked. Political initiatives might include import controls on both foreign cars and components; government support for smaller manufacturers to invest in new technology and new products; abolition of the UK car tax *for domestically manufactured cars with at least 80 per cent local component content* (which would stimulate domestic production only, rather than suck in imports); a greater price differential between petrol and diesel fuel to encourage the domestic production of diesel passenger cars[13]; and the imposition of an effective (i.e. non-cost based) 80 per cent component content rule for 'British' cars manufactured in the UK or imported into the UK by UK-based motor manufacturers such as Ford, GM, and Peugeot. Given the geographical concentration of the industry in the West Midlands, national policies should be supported at local level through investment support for manufacturers to move from low technology production to the production of higher value added components and subassemblies, support for investment and diversification into electrical and electronic components, and support for training, export strategies, and marketing advice for smaller companies provided on a bureau basis. In the absence of these or similar initiatives to protect the interests of labour and the components industry in general, the adoption of just-in-time may yet transpire to be just-too-late to secure the long-term future of the ailing UK automotive industry.

Notes

1. This figure is based on a recent survey by PEIDA Consultants for the Society of Motor Manufacturers and Traders (SMMT), and includes both direct employment at the vehicle and component manufacturers and 'indirect' employment among the suppliers of these companies. There is also significant employment in the motor trade (i.e. dealerships and garages).
2. The component content of cars is generally measured by ex-works costs which allows non-manufacturing costs such as marketing expenses, distribution costs, servicing and guarantee costs, direct and indirect labour costs, and even profits to be included in the calculation, thus distorting the 'true' level of locally sourced components. Thus, Ford and GM/Vauxhall can claim a component content over 60 per cent and 50 per cent respectively (HCTIC, 21 January 1987:49, Ref. 143–i and 28 January 1987:85, Ref. 143–ii). The accepted 'minimum' content for a vehicle to qualify as 'British' is 60 per cent, but there is no legal imposition of such a 'content rule'. Nevertheless, the rule can be effective in the important fleet car market as many companies operate a 'Buy British' policy (the 60 per cent content level originated from a EEC/EFTA agreement, but no specific content percentage is required by EEC Council Regulation 802/68, June 1968, which defines EEC origin criteria. The HCTIC is currently considering legal imposition of local content levels and whether or not this would impinge on EEC regulations).
3. Because of its dependence on a declining UK vehicle industry, the UK car components industry suffered the biggest collapse of any car producing nation in the world (Jones, 1985:9). The UK has suffered both the longest and the biggest decline of vehicle production – in 1984, for example, UK car production was just 47 per cent of its post-war peak achieved as long ago as 1972, while for CVs the comparable figure was only 48 per cent of peak production realized in 1969. All other major motor manufacturing nations have maintained production levels at or above 70 per cent of their post-war peak for both

cars and CVs (in some cases production has declined by less than 10 per cent).

4. West German manufacturers such as Bosch have a firm hold on many of these high-tech products, based on the strength of their domestic motor industry (West Germany not only has a total production level four times greater than that in the UK but also produces 13–14 times as many high performance, high value cars, HCTIC, 25 February 1987:182, Ref. 143–vi and 6 May 1987:298, Ref. 143–xii).

5. The correlation coefficient between employment decline in the vehicles and engines industry and the components industry during the period 1974–85 was $r^2 = 0.99$, calculated from Department of Employment figures (HCTIC, 4 February 1987:110, Ref. 143–iii).

6. Overall the (smaller) specialist RE component manufacturers have been the more profitable group in recent years (see ICC Reports), but they themselves now recognize their limited long-term prospects in a UK market characterized by declining domestic production, increasing import penetration, and greater competition in the aftermarket (see, for example, the evidence of Avon Tyres, HCTIC, 18 March 1987:27, Ref. 143–viii). As a result, a number of RE manufacturers have formed the 'Impact Group' in an attempt to pool their resources and win more export orders. The government has also passed legislation (Autumn 1982) to secure access for this group to the franchized sector of the aftermarket where vehicle manufacturers and vehicle importers had effective (monopoly) control over replacement equipment. However, opening up this segment of the market to greater competition appears to have opened the door for foreign component manufacturers to secure greater access to the British sector of the market.

7. Up to 60 per cent of the cost of a car comes from bought-in components and, therefore, motor manufacturers have made component prices a prime target for cost reduction (see *Financial Times*, 26 May 1983 and Flax, 1983).

8. Before the abolition of franchise agreements in August 1982, which protected the aftermarket share of vehicle manufacturers, the UK vehicle manufacturers and their subsidiary component suppliers – Unipart/BL, Motorcraft/Ford. AC Delco/GM–Vauxhall, and Motoquip/Peugeot–Talbot – held around 80–90 per cent of the replacement market for their own vehicles (largely controlled by franchise arrangements) and around 25 per cent of the total aftermarket for replacement equipment (MMC, 1982:6).

9. Typical Western 'lot-size' or 'order-point' systems of production/inventory control are described as 'push-through' approaches as raw materials, parts and subassemblies are purchased or produced and 'pushed' into work-in-progress inventories even when they are not currently needed by the next stage of the production process (see Rice and Yoshikawa, 1982).

10. Western production control systems are based on the two interrelated principles of economic order quantities (EOQ) and 'buffer stocks'. At its simplest, the EOQ is defined as that quantity of production per shift, cycle or set-up that achieves the best trade-off or balance between set-up costs and the costs of holding stocks (thus as set-up costs increase batch size increases, and as handling, storage and carrying costs increase batch sizes are reduced). This principle is combined with the practice of staging buffer stocks between successive work stations to keep production going in the event of downtime at any individual work station (each work station simply works into its buffer

stocks until production is restarted). JIT, in contrast, is based on the principles of reducing set-up times to zero and holding no buffer stocks. Theoretically, therefore, the optimum lot size can be reduced all the way down to one unit (see Rice and Yoshikawa, 1982; Schonberger and Schniederjans, 1984; and Abegglen and Stalk, 1985:95).

11. This particular plant is essentially a machining shop, and unlike many other ACC automotive component factories there is relatively little fabrication/assembly work. However, while the specific form of production reorganization will vary across plants, products, and processes, the organizational principles are generic rather than specific.

12. The introduction of JIT at this plant contrasts sharply with the recent experience of other companies. Lucas Electrical, for example, threatened its workforce with factory closure and/or individual dismissal if they did not accept new practices associated with a JIT system (see Turnbull, 1986), and Bedford Vans has recently adopted a similar line in establishing a joint venture with the Japanese manufacturer Isuzu (see *Guardian*, 2 July 1987 and 7 July 1987).

13. The diesel passenger car market is the most rapidly expanding segment of the automotive market – European sales of diesel cars accounted for 5.3 per cent of total European car sales in 1979, but well over 15 per cent in 1985. Britain has lagged behind the rest of Europe in this market, with diesel sales accounting for only 2.5 per cent of total registration in 1984 compared with over 13 per cent in both France and West Germany and over 26 per cent in Italy.

Bibliography

Abegglen, J.C. and G. Stalk. 1985. *Kaisha: The Japanese Corporation*. New York: Basic Books.

Abernathy, W.J., K.B. Clark and A.M. Kantrow. 1983. *Industrial Renaissance: Producing a Competitive Future for America*. New York: Basic Books.

Aggarwal, S.C. 1985. 'MRP, JIT, OPT, FMS? Making Sense of Production Operations Systems', *Harvard Business Review*, September–October, 8–16.

Ballance, R.H. and S.W. Sinclair. 1983. *Collapse and Survival: Industry Strategies in a Changing World*. London: Allen & Unwin.

Bessant, J., D.T. Jones, R.L. Lamming and A. Pollard. 1984. 'The West Midlands Automobile Components Industry: Recent Changes and Future Prospects', West Midlands County Council, Economic Development Unit Sector Report No. 4.

Birmingham Economic Policy Group. 1987. 'Visit to Austin Rover', Motor Industry Local Authority Network, EPPR No. 1.

Business Week, 1985, 14 May.

Celley, A.F., W.H. Clegg, A.W. Smith and M.A. Vonderembse 1986. 'Implementation of JIT Concepts and Practices in the United States', Business Research Centre, College of Business Administration, University of Toledo, W.P. No. 86–5.

Cook, J. 1984. 'Kanban, American-Style', *Forbes*. October, 66–70.

Crosby, L.B. 1984. 'The Just-In-Time Manufacturing Process: Control of Quality and Quantity', *Production & Inventory Management*. Fourth Quarter, 21–33.

160 *Restructuring in Automotive Components*

Dodwell. 1986. *The Structure of the Japanese Auto Parts Industry*. 3rd edn, Dodwell Marketing Consultants.

Dohse, K., U. Jürgens, and T. Malch, 1985. 'From "Fordism" to "Toyotoism"? The Social Organization of the Labor Process in the Japanese Automobile Industry', *Politics and Society*. Vol. 14, No. 2, 115–46.

Engineer, various dates.

Engineering Computers. 1986. 'Just-In-Time Improves the Bottom Line', September, 55–60.

Financial Times, various dates.

Finch, B. 1986. 'Japanese Management Techniques in Small Manufacturing Companies: A Strategy for Implementation', *Production & Inventory Management*. Third Quarter, 30–8.

Flax, S. 1983. 'A Hard Road for Auto Parts Makers', *Fortune*. March, 108–13.

George, M. and H. Levie. 1984. *Japanese Competition and the British Workplace*. London: CAITS.

Grunberg, L. 1986. 'Workplace Relations in the Economic Crisis: A Comparison of a British and a French Automobile Plant', *Sociology*, Vol. 20, No. 4, 503–29.

Guardian, various dates.

Hayes, R.H. and S.C. Wheelwright. 1984. *Restoring Our Competitive Edge*. New York: Wiley.

HCTIC. 1987. 'Minutes of Evidence Taken Before the House of Commons Trade and Industry Committee, The UK Motor Components Industry'. London: HMSO.

Ichiyo, M. 1984. 'Class Struggle on the Shopfloor – The Japanese Case (1945–84)', Ampo: *Japan-Asia Quarterly Review*. Vol. 16, No. 3, 38–49.

ICC. Various editions. *Business Ratios: Motor Components and Accessory Manufacturers*, Inter Company Comparisons.

Incomes Data Services. 1986. 'Flexibility At Work', Study 360, April.

Investors Chronicle. 1982. 22 January.

Jones, D.T. 1985. 'The Import Threat to the UK Car Industry', University of Sussex: SPRU.

Kamata, S. 1982. *Japan in the Passing Lane*. London: Allen & Unwin.

Monden, Y. 1981. 'What Makes the Toyota Production System Really Tick?', *Industrial Engineering*. January, 36–46.

Monopolies and Mergers Commission. 1982. *Car Parts*. London: HMSO.

Pendlebury, J. 1984. 'West Discovers Just-In-Time', *Chief Executive*. Vol. 17, March, 25–6.

Price Commission. 1979. *Prices, Costs and Margins in the Manufacture and Distribution of Car Parts*. London: HMSO.

Rhys, G. 1985. 'The Economic Outlook for the Motor Industry', SMMT Conference on The Future of the Motor Industry in the Midlands, Birmingham, 3 April.

Rice, J.W. and T. Yoshikawa. 1982. 'A Comparison of Kanban and MRP Concepts for the Control of Repetitive Manufacturing Systems', *Production & Inventory Management*. First Quarter, 1–13.

Saipe, A.L. 1984. 'Just-In-Time Holds Promise for Manufacturing Productivity', *Cost & Management*. Vol. 58, May–June, 41–3.

——. and R.J. Schonberger. 1984. 'Don't Ignore Just-In-Time Production', *Business Quarterly*. Vol. 52, Spring, 60–6.

Sayer, A. 1986. 'New Developments in Manufacturing: The Just-In-Time System', *Capital and Class*, Winter, No. 30, 43–72.

Schonberger, R.J. 1982. *Japanese Manufacturing Techniques: Nine Hidden Lessons in Simplicity*. New York: Free Press.

——. 1983. 'Japanese Manufacturing Techniques: Nine Hidden Lessons in Simplicity', *Operations Management Review*, Spring, 13–18.

——. and M.J. Schniederjans. 1984. 'Reinventing Inventory Control', *Interfaces*. Vol. 14, No. 3, 76–83.

Shaiken, H., S. Herzenberg and S. Kuhn. 1986. 'The Work Process Under More Flexible Production', *Industrial Relations*. Vol. 25, No. 2, 167–83.

Society of Motor Manufacturers and Traders. 1985, 1986. *Motor Industry of Great Britain: World Automotive Statistics*. London: SMMT.

Spurgeon, E.V. 1983. 'Batches of One – The Ultimate in Flexibility', *Production Engineering*. September, 50–2.

Tailby, S. and P.J. Turnbull, 1987. 'Learning to Manage Just-In-Time', *Personnel Management*. January, 16–19.

Turnbull, P.J. 1986. 'The "Japanisation" of Production and Industrial Relations at Lucas Electrical', *Industrial Relations Journal*. Vol. 17, No. 3, 193–206.

——. 1988. 'The Limits to Japanisation: "Just-In-Time", Labour Relations and the UK Automotive Industry', *New Technology, Work and Employment*. Vol. 3, No. 1, 7–20.

Willman, P. and G. Winch, 1985. *Innovation and Management Control: Labour Relations at BL Cars*. Cambridge: Cambridge University Press.

Wood, S. 1986. 'Technological Change and the Cooperative Labour Strategy', EGOS Autonomous Working Group on Trade Union Research, Conference on Trade Unions, New Technology, and Industrial Democracy, University of Warwick, 6–8 June 1986.

Fork Lift Trucks

Continuing the theme of new production systems and the industrial structures which support them, this study analyses one firm's failure to establish the conditions for successful restructuring.

The industry is potentially important as a producer of automated handling systems which could enhance manufacturing productivity. The study examines the structural constraints preventing British firms from establishing a presence in this advanced sector.

The firm, a subsidiary of a state-owned corporation in the 1970s, was used as a focus for the centralization of capital ownership in the fork lift truck industry and attempts to develop mass-production. The failure of this approach, and its results in transforming a once highly profitable firm into a loss-making enterprise, are analysed.

Privatized in 1981, the firm sought to restore its previous advantage in customized production and to reduce unit costs through the introduction of 'just-in-time' (JIT) manufacturing methods. Management proved unable to reorganize production effectively, however, because of the firm's weak competitive position within the industry, forcing it into the production of one-offs to secure sales orders. This in turn weakened its position in relation to supplier companies, the latter being under pressure from larger motor vehicle manufacturers to ensure JIT deliveries.

The consequent disorder in production enabled the trade unions locally to maintain their presence and temporarily defend members' interests. Co-operation with management in production served to maintain output, but by itself could not arrest the company's decline. The study argues that, separately, neither the union's political influence within industrial planning, nor its undoubted shopfloor strength were sufficient to influence substantive outcomes.

6

Handling Decline in the UK Fork Lift Truck Industry: A Case Study

Stephanie Tailby

Introduction

This chapter examines the complexities of capital and labour restructuring in the fork lift truck industry. It focuses on the specific circumstances of one British company and the reasons for this firm's inability to adjust to the changing competitive conditions within the industry over the period from the mid-1970s. At the start of this period LiftCo had been identified as one of the leading, British-owned fork lift truck producers, a highly profitable company operating in an industry ranking as one of the more successful sectors of UK engineering. Rising levels of import penetration over the late 1970s reflect the industry's difficulties in responding to intensified international competition. LiftCo's reversal of fortunes, however, has been exceptionally dramatic. A loss-making concern at the end of the 1970s, attempts to return it to profitability were unsuccessful. LiftCo traded on the verge of bankruptcy through to the middle of the 1980s.

The firm's relative decline within the sector is explained by an analysis of the reasons why the agencies positioned to promote its development proved to be ineffective. Since LiftCo had been a subsidiary of the publicly-owned British Leyland motor company up until 1981, these agencies include the state and management in their interaction with labour and the trade unions.

Under British Leyland, the company was distinguished from its competitors within this sector by its vulnerability to the shifting perspectives on

restructuring pursued by successive Labour and Conservative governments. The fork lift truck industry was included in the 1974–9 Labour
government's 'industrial strategy'. Indicative planning was aimed at promoting the concentration of production in the British-owned sector and
higher levels of productivity through the pursuit of manufacturing scale
economies. Party to these initiatives, the ability of the trade unions to
influence industrial development was undermined, in part by fragmented
labour organization. But indicative planning proved inadequate to the task
of co-ordinating industrial change and restructuring. Thus, while LiftCo's
growth, through acquisitions, played an important part in the partial
restructuring of capital ownership resulting from these initiatives, private
companies opted to pursue a variety of individual responses to squeezed
profit margins and intense price competition in the volume sector of lift
truck markets.

With developing overcapacity in the industry, and the political transition
in Britain towards market-led restructuring, support for LiftCo's expansion and modernization was withdrawn in the late 1970s. The company was
returned to the private sector in 1981 with employment and capacity
severely reduced. Under new ownership, the management attempted to
restore profitability through the reorganization of production to secure
'economies of scope'. Involved was an adaptation of the just-in-time (JIT)
production and inventory control system employed by large-scale Japanese
competitors to attain minimum unit costs in the manufacture of standardized models for the volume sector of Western lift truck markets. Representing an adaptation of conventional mass production technology, this
system embraces changes in the organization of the manufacturing labour
process and in relationships with component supplier firms which allow
companies to assemble a more diversified range of models without sacrificing economies of standardization. JIT can also be employed to bring
short-batch production to more closely approximate a repetitive flow-line
assembly process. At LiftCo, the aim was to reduce unit costs and expand
sales through product diversification and the development of specialized,
less price-competitive market niches.

Classified as a form of flexible manufacturing, and viewed as a route
through which firms can attain higher levels of productivity and a greater
responsiveness to market opportunities, the JIT system has been identified
by some commentators as a managerial solution to the crisis of profitability
in a wide range of industrial sectors. Its effective operation, however, is
dependent upon the firm's ability to restructure internal labour–management relations and relations with external supplier companies. By virtue of
its weak competitive position in overcapitalized product markets, LiftCo
was forced to accept an increasingly irregular sales order intake. This
inhibited management's ability to restructure relations with suppliers and
secure 'JIT' parts deliveries, and with this to reduce unit costs through
faster throughout in production.

The persistence of parts shortages and bottlenecks in production emphasized management's dependence upon labour's adaptability and co-operation in production to maintain output. Exacted in the context of internal
job losses and unemployment in the external labour market, labour's co-
operation nevertheless proved insufficient to support the firm's recovery.
A wider set of relationships, largely outside management's control, combined to frustrate the restoration of profitable production.

The analysis is developed in four sections. The first examines the forces
for change in the structure of the British lift truck industry and in the
organization of production at the level of the enterprise experienced from
the 1970s. The second section examines trade union organization in this
industry and the character of workplace labour relations at the case study
company. The reorganization of the company under British Leyland is
discussed in the third section, and management's ultimately unsuccessful
attempts to restructure production following the privatization of LiftCo in
1981 are considered in the final section.[1]

The Fork Lift Truck Industry

Located within the capital goods sector of the economy, the pace of the
fork lift truck industry's development is heavily dependent upon industrial
growth as a whole. The industry in the UK enjoyed a period of sustained
expansion up to the 1970s, fuelled by generally high levels of profitability.[2]
The context was an accelerating rate of growth of demand as firms sought
to raise output and productivity through the substitution of powered
equipment for labour in the internal transport of raw materials, inventories
of work-in-progress and final products. Between 1963 and 1973, the output
of basic truck units increased by 300 per cent in comparison with an
increase of 50 per cent for UK industrial output as a whole (*Guardian*, 18
August 1982).

The manufacture of fork lift trucks in the UK dates from the immediate
post-Second World War period. The four US-owned multinationals that
were to dominate world markets up until the 1970s established branch
plants in Britain.[3] A larger number of smaller, British-owned companies
entered the industry during the 1950s and 1960s. LiftCo, established
around the turn of the century to manufacture 'bespoke' engines for the
nascent motor vehicle trade, diversified into the lift truck industry in the
1940s. Via a series of ownership changes dating from the 1960s, the firm
subsequently became part of British Leyland. Classified as one of the
leading British-owned producers, this firm was reported to be one of the
most consistent profit-makers in the industry in the mid-1970s.

Inward US investment, in conjunction with the expansion of a number
of British-owned companies, had established the UK as 'one of the most
important centres in Europe' for the production of fork lift trucks in the

post-war period to the 1970s (*Financial Times*, 2 May 1980). While the UK market remained relatively small, in world terms,[4] the British industry had been able to expand through exports with the opening-up of markets in Europe and in Commonwealth countries. Up until the early 1970s the UK had ranked second to the United States within the OECD in terms of export sales of these vehicles.

With the development of this industry in other West European economies, however, and the entrance of new producers, the British industry experienced intense international competition for the first time from the beginning of the 1970s. By the middle of the decade the industry in the UK had slipped from second to fourth position within the OECD export league, behind the US, West Germany and Japan. Import penetration of the UK market had also risen to 25 per cent of domestic sales. At this stage, US and West German companies accounted for the bulk of UK import sales, with the former also bringing in imports for their UK-based subsidiaries. West German trucks were reported to be more highly engineered than similarly-priced British models, with technically-improved designs resulting in improved vehicle performance and reliability in service (*Financial Times*, 2 May 1980). West German and US-owned companies, however, were in turn experiencing intense price competition from Japan where the lift truck industry had been rapidly developed from the 1960s.

High profit levels had fuelled the expansion of this sector in the UK and internationally. By the late 1970s, however, the fork lift truck industry was reported to be operating with 30 per cent excess capacity on a world-wide scale (NEDO, 1978). Intensified competition generated the continuing build-up of capacity, while the rate of growth of demand had begun to slacken with the deepening of the economic crisis. Product markets collapsed at the turn of the 1980s as end-user industries plunged into recession and, with stagnant output growth, buyers attempted to maintain profitability principally through rationalization or disinvestment. World output of fork lift trucks peaked at 390,000 units in 1978, but had fallen to 298,000 by 1982 (*Financial Times*, 25 February 1985). In the UK, sales of lift trucks fell from approximately 12,000 units in 1980, to 7,000 in 1981 (*Guardian*, 18 August 1982). But while British, West European and US-owned fork lift truck manufacturing companies experienced the erosion of profitability with the contraction of demand, Japanese firms continued to make advances in Western lift truck markets.

The mechanization of materials handling activities in the post-war period had formed a key element in the 'Fordist' flow-line mass assembly production system, adopted in a range of consumption goods sectors. While the production of fork lift trucks is a relatively straightforward fabrication and assembly operation, however, manufacturing economies of scale and fast throughput production techniques had been of limited significance in terms of the nature of competition in the period to the

1970s. Firms had been able to expand through product diversification, adding new product lines and building vehicles of different sizes, lifting capacities and with a range of special fixtures for different lifting applications in order to develop profitable market openings. Hence, outside the centrally-planned economies,[5] firms were principally engaged in short-batch assembly.

Nevertheless, the US multinationals had been able to exploit scale economies in the manufacture of components through central (or global) sourcing strategies. Bought-in components and raw materials represented the major item in manufacturers' production costs – between 60 and 65 per cent of the total. Overheads (fixed plant, equipment and indirect production employment) represented the second major expense at 30 per cent, with direct labour (in fabrication and assembly) forming only 5 per cent of the total (Abegglen and Stalk, 1985:81; NEDO, 1978). A high ratio of indirect to direct production workers reflected the complexity in manufacturing, with ancillary progress chasers and production schedulers required to support and co-ordinate the various lines produced.

British companies outsourced the bulk of their requirements for components and fabricated subassemblies. Higher levels of integration – and lower piece part costs – were precluded by the scale of production and range of models manufactured. Thus, and in contrast with developments in other Western European economies, capital ownership in the British lift truck industry remained relatively fragmented in the 1970s, with the British-owned sector embracing a high proportion of very small-scale, specialist companies producing for discrete niches of the market. With the exception of Lansing Bagnall, which had captured a 20 per cent share of the UK market (*Materials Handling News*, October 1976:4), the larger British firms would be classified as medium-sized companies. LiftCo, for example, one of the largest British companies, had the capacity to build 3000 trucks per annum at its single manufacturing site in the West Midlands. While this firm was a subsidiary of a larger parent organization, the British industry comprised a high proportion of private limited companies, owned and controlled by the founding families.

This industrial structure had been supported by favourable demand conditions. British firms competed in the 'volume' sector of the market, embracing trucks in the low-lifting capacity range and representing some 70 per cent of UK sales volume, and the principal product lines for export markets. They also manufactured a more extensive product range, although less comprehensive than that offered by the US multinationals. To a greater extent than the latter, however, British firms engaged in 'customized' production, modifying basic designs to end-user specification and manufacturing 'one-offs'. Customized models, in common with larger vehicles and trucks for specialist lifting applications, command a high sales price. Hence, through their responsiveness to customer demands and enterprise in developing specialized market niches, British firms had enjoyed a period of considerable prosperity up until the mid-1970s and in the context of generally buoyant market conditions.

Representatives of the British companies argued that flexibility in meeting the demands of the customer was essential to secure most sales, and that this in turn meant that there were limited economies of scale in producing trucks in vast quantities, because of the large number of 'specials' and wide product ranges (*Materials Handling News*, February 1977:21). Nevertheless, the export success of the Japanese industry has been based on the high volume manufacture of low-priced, standardized models in the low-lifting capacity, or volume range. The industry in Japan is dominated by five companies, all large-scale producers and subsidiaries of larger conglomerate corporations two of which – Toyota and Nissan – have investments throughout the wider automotive sector. In the mid-1970s, the output of the Japanese industry totalled 87,000 units, which compared with a UK output of 28,000 (*Materials Handling News*, April 1976:16).

Entering international competition at a relatively late stage, Japanese firms had acted to expand export sales by focusing production for the volume sector of the market and organizing production to secure the levels of productivity associated with mass-production. Through focused production alone, for example, Toyota is estimated to have achieved a 21 per cent cost advantage over a US competitor's similarly-sized manufacturing facility in West Germany geared to the production of 20 separate truck 'families' (Abegglen and Stalk, 1985:85). Japanese companies attained additional improvements in productivity by adapting US designs, with models in the low-lifting capacity range modified to reduce the labour time involved in the fabrication and assembly of each unit, by employing new methods of inventory control and new forms of work organization, and by investing in advanced technology with the use of robotics in the fabrication of subassemblies. Japanese import penetration of the UK market rose from 12 to 25 per cent over the period 1979–81 (*Guardian*, 18 August 1982), and Japanese firms made similar inroads into the US domestic market as the 'depressed economic climate forced operators to opt for trucks of a basic workmanlike design to help cut costs and reduce overheads' (*Business & Market Research*, 1981:13).

Japanese firms have also been at the forefront in the development of new, automated materials handling systems (AHS), or mechanical handling hardware controlled by computer and embracing automated guided vehicles (driverless fork lift trucks) and automated storage and retrieval warehousing systems (*Materials Handling News*, April 1976:16 NEDO, 1986). Advanced materials handling technology represents an important source of productivity growth for a wide range of consumption and service goods industries. Employed in conjunction with computer-aided-design and manufacturing (CAD/CAM), the new technology enhances management's ability to control and co-ordinate complex production systems and allows firms to reap scale economies within short-batch production (NEDO, 1986). For manufacturers of materials handling equipment, AHS technology represents the growth area of international markets. Thus, while firms face long-term stagnation in the rate of growth of demand for conventional (mechanical) handling technologies in the advanced industrial economies, reports predict an annual rate of growth of 12 to 15 per cent in world markets for AHS technology through to the 1990s (NEDO, 1983).

Nevertheless, the actual rate of growth of demand is dependent upon the pace of change in end-user industries and in the UK in the mid-1980s the market for AHS remained relatively underdeveloped, with 'many users . . . unaware of, or unresponsive to, the potential advantages of systems technology' (NEDO, 1986:9). While the relative cost of the new technology has inhibited its application in the UK, the scale of investment required in the development and manufacture of advanced handling systems (with R&D expenditure estimated to be five times higher in comparison to conventional technology) has constrained British lift truck companies from entering this growth area of world markets (NEDO, 1983). Thus, while British manufacturers had displayed a considerable responsiveness to established market opportunities, by the late 1970s they were being undercut by Japanese producers in the volume sector of lift truck markets and were lagging behind their Japanese, West German and US competitors in terms of product innovation and the development of new market openings (NEDO, 1983).

A series of mergers and acquisitions in the mid-to-late 1970s served partially to reduce the fragmented structure of capital ownership in the British fork lift truck industry. This reorganization of capital ownership was 'prompted partly' by the industry's NEDO Sector Working Party, established under the 1974–9 Labour government's 'industrial strategy' (*Financial Times*, 2 May 1980). As one of the most profitable British companies, and a subsidiary of British Leyland, LiftCo occupied an important role in this attempt to accelerate the process of restructuring via indicative planning. The trade unions were involved in these initiatives and hence before the outcomes are discussed, the organization of labour in the industry and the nature of workplace industrial relations at the case study company are considered in the following section.

Labour Organization and Workplace Labour Relations

In employment terms, the fork lift truck industry is a relatively small sector of UK engineering. Aggregate employment was estimated at 14,000 in 1978, with managerial, administrative, technical and clerical grades representing 50 per cent of the total (NEDO, 1978). The fragmentation of the workforce, divided between some 20 companies, was held by some managers to have had a beneficial influence on labour relations. Thus, the small scale of plants was argued to account for the absence of labour militancy and the legacy of 'good industrial relations' enjoyed by firms in the industry (*Materials Handling News*, February 1977:21).

Workers in the industry are represented by a range of general, staff and craft sections of general unions, with multi-unionism a feature of labour relations at the enterprise level. The UK industry embraces a small proportion of non-union firms.[6] Collective bargaining structures are decentralized with wage bargaining at factory level.

LiftCo had 1500 employees in 1975, with the production workforce representing about half of this total. Staff and hourly-paid workers were unionized with clerical, supervisory and technical grades represented by APEX, ASTMS and TASS respectively. Union density on the production workforce side was 100 per cent, with membership divided roughly equally between the AUEW and TGWU. This had followed an active recruitment drive since, in the 1960s, only skilled, direct production workers were organized. A central, joint shop stewards' committee had been established by the mid-1970s, and it negotiated base rates for comparable grades across the hourly-paid workforce. While British Leyland had attempted to extend measured daywork across its operations, LiftCo retained its piece-work payment systems to the end of the 1970s. Hence, the principal focus of steward activity remained sectional wage bargaining.

The production workforce was spatially decentralized. The firm's manufacturing departments were dispersed, reflecting its expansion in an urban centre where industrial sites were at a premium. Operators in the machine shops were on an individually-based piecework scheme. The fabrication department, located on the same site, manufactured masts for the truck assembly plant, with other subassemblies outsourced. Workers in this area were engaged in short-batch production with masts manufactured to various specifications and sizes for the different models produced. The work progressed from the flame-cutting area, through welding, machining and mast assembly.

Masts and machined components were transported across town to the truck assembly plant. The company had five main product lines with vehicles assembled over a series of fixed work stations, arranged in track formation although connected by overhead cranes rather than by moving conveyor belts. Each of these lines was assembled on a separate track by a gang of fitters. Larger vehicles and 'true specials' were assembled in bays to either side of the tracks. From assembly, the trucks were driven into the paint shop and thereafter to the test and final-fitting departments.

The fitters in assembly retained their skilled status through the 1970s and were recruited for their general skills and experience in work of a non-repetitive kind. It was on this basis that they maintained a differential with production workers in car assembly plants in the district and thus were relatively well placed within the district's engineering earnings structure. The fitters were on a gang bonus scheme, which served to foster the development of a cohesive group of workers in this area. Job controls restricted management's ability unilaterally to determine the organization of production, although the gang bonus system also supported a measure of flexibility in the utilization of labour in this area with shopfloor supervision able to reallocate labour according to sales order intake and production schedules.

The firm's inclusion within British Leyland, according to management representatives, had brought to an end an era of close or harmonious labour–management relations. The workforce had apparently become 'less compliant' and stewards more assertive in their negotiations with the company. The period from the mid-1970s was marked by an increased wage militancy as internal relativities became distorted and the various areas pressed to re-establish customary differentials. Nevertheless, management at LiftCo were as reluctant as the unions to discard piecework, and had successfully resisted British Leyland's attempts to extend measured daywork to this subsidiary. An earlier experience of flat-rate payments had apparently convinced the company that measured daywork could have only deleterious consequences for productivity, defined in terms of the intensity of work effort.

In practice, the stewards at LiftCo remained relatively isolated from their counterparts in British Leyland's car divisions, electing instead to form closer links with union representatives from other firms included within the Special Products Division. Management's ability to resist the introduction of measured daywork, moreover, reflected LiftCo's status within British Leyland as a peripheral – although initially a highly profitable – part of the parent corporation. This in turn is in contrast with the close direction exercised by firms such as Toyota in Japan over the activities and development of their lift truck subsidiaries.

Rising wage costs do not provide a convincing explanation for LiftCo's deteriorating performance in the late 1970s, given the cost structure of firms within the fork lift truck industry. While direct production workers remained relatively well placed within the district earnings structure, and relatively well paid in comparison with comparable grades elsewhere in the UK lift truck industry, indirect production workers and clerical grades (mainly women) were far less favourably positioned. The company maintained one of the largest differentials between hourly-paid and clerical earnings in the district. British companies, moreover, have been challenged by large-scale, high productivity Japanese firms and not by low-wage producers. Wage militancy might be expected to stimulate change and investment in labour-displacing technologies. Its general absence across the UK industry had in practice served to maintain the industry's fragmented structure, and the survival of small-scale, marginal producers.

With the economic downturn of the mid-1970s in Britain, a number of these firms had been exposed to the possibility of bankruptcy and take-over. State policy in relation to the industry in this period was aimed at accelerating the process of restructuring and the concentration of production via indicative planning, the sector's deteriorating export performance meriting its inclusion within the Labour government's 'industrial strategy'. The trade unions were party to these initiatives, although their ability to influence industrial change through participation in 'planning' was undermined both by fragmented labour organization and the commitment to co-operation with the employers who were, in turn, divided by competition. In effect leaving the initiative for change with the employers, indicative planning therefore failed substantially to ameliorate the constraints to autonomous restructuring which, in addition to the competitive relations among firms, included the sector's dependence upon industrial development more generally. The weakness of the government's commitment to planning is most vividly exemplified by LiftCo's increasing difficulties. As a subsidiary of a state-owned corporation, this company afforded the potential for strategic intervention to promote industrial change and development. The assistance it actually achieved, however, was limited.

The Politics of Change

Indicative Planning

Under the auspices of the NEDO Industrial Trucks Sector Working Party (SWP), employers and trade union representatives were brought together in an attempt to secure commitment to a 'planned' reorganization of the fork lift truck industry. All major firms in the UK industry were represented in the SWP's discussions. British-owned companies were already included within a single trade association – the British Industrial Trucks Association. In contrast, with decentralized structures of collective bargaining in the industry, union (or in some cases, employee) representatives were brought together on an industrial basis for the first time. The active involvement of shopfloor representatives in the planning process was limited, however, with national officials taking the lead in the SWP's discussions.

Given the small size of the sector in terms of employment, and diversity of union representation within it, the national officials can have had only a limited knowledge of the industry and its particular difficulties. Nevertheless, the unions were committed at a national level to tripartism and to the need for industrial change in order to protect jobs, with representatives on the SWP taking very seriously the 'threat' of Japanese competition. But the problem, as they perceived it, was the rivalry between firms which meant that 'they are all wanting to maximise their profit margins, and hence they will not declare their hand on the policies they are going to adopt' (TUC Report, 1977:67).

The SWP had limited powers. Its establishment represented an attempt to encourage voluntary restructuring, but without the backing of finance to induce change through the promise of subsidy. Discussion focused on the British industry's performance in export markets and the need to reduce import penetration of the domestic market. Recommendations for action were principally directed at the British-owned sector of the industry. Thus, while US-owned subsidiaries accounted for 50 per cent of domestic production, the need to encourage existing US investment – rather than to co-ordinate the development of foreign and domestic capital – was emphasized. The institutions of 'planning', moreover, reflected the limited aims of the government's industrial strategy. Separate Sector Working Parties were established for the fork lift truck industry (Industrial Trucks) and for the wider mechanical handling sector. This served to preserve traditional subsectoral divisions which abroad were becoming obsolete with the move into the manufacture of automated materials handling equipment (NEDO, 1983:7). The narrowly sectoral focus of planning in the UK, however, failed to provide the stimulus to product innovation, which would necessarily have required an approach co-ordinating the development of the materials handling industry and its end-user markets in consumption and service goods sectors.

The SWP's reports acknowledged the possibility of expanded capacity on a world-wide scale being 30 per cent underutilized by the end of the 1970s and, in this context, the likelihood of product markets becoming increasingly competitive. Nevertheless, fairly optimistic growth targets were set for the British industry in both UK and export markets. In order to achieve these targets, British firms were urged to engage in horizontal amalgamation. The concentration of production would enable firms remaining in competition to rationalize facilities and product lines, and thereby secure manufacturing scale economies in vehicle assembly and in the sourcing of components, with standardization permitting higher levels of vertical integration.

Anxious to expand and to increase market share, the larger British firms could agree on the need for mergers and acquisitions. But while they acknowledged the benefits of increased firm size, in terms of the 'marketing' economies that might thereby be derived, employers expressed considerable scepticism as to the advantages to be gained from large-scale production and standardization (*Materials Handling News*, February 1977:21). Through product diversification, individual companies had been able to maintain high profit margins by developing specialized, less price-competitive niches of the market, in spite of developing overcapacity in the industry and intensified foreign competition in the volume market. The volume production of trucks in the low-lifting capacity range, on the other hand, was likely to bring them more directly into competition with larger Japanese and US-owned corporations. Thus, and in response to Japanese competition, US multinationals were moving into the mass production of trucks for the volume market either, as in the case of Clark Equipment, by reducing the number of product lines produced, and concentrating production on larger sites, or, like Hyster, by opening new facilities equipped with advanced technology[7] (*Materials Handling News*, May 1976:4).

In 1977, the Industrial Trucks Sector Working Party reported that, 'action towards restructuring had largely been taken by the industry itself' (cited in *Materials Handling News*, July 1977:19). Certainly there had been a series of takeovers in the UK industry; the number of British firms was reduced to 16, with three major British employers emerging. In practice, however, private companies had pursued a variety of individual, competitive strategies. Lancer Boss, for example, had eschewed involvement in the merger activity, with its owners declaring that the firm's intention was to 'stay independent and private and a specialist' (*Materials Handling News*, August 1978:4). Reorienting production to evade intense price competition through the manufacture of larger vehicles and side-loading equipment, this firm alone out of the British companies remained profitable through the recession of the early 1980s. Lansing Bagnall had acquired two currently profitable companies, giving it a 30 per cent share of the UK market. These acquisitions complemented the firm's main electric truck range, giving it new market opportunities, in particular in the expanding sales area of rough terrain vehicles for agricultural end-markets. Through a strategy of product specialization on different sites, moveover, the company has to a certain extent been able to combine the benefits of volume and product variety.

LiftCo, in contrast, had acquired two financially weak subsidiaries, one of which duplicated the firm's existing product range. With financial control exercised by British Leyland, the company had proved more amenable to direction. Thus, the original intention was for LiftCo to become a volume producer of a more standard range of trucks in the low-lifting capacity range. British Leyland had been oriented along an expansionary path following the recommendations of the Ryder Report, which expressed the Labour government's commitment to support the corporation's long-term viability through investment and modernization. As with the parent company, however, LiftCo's subsequent difficulties partly reflected the absence of a coherent and consistent strategy for the industry in which it operated. LiftCo, moreover, suffered from a lack of investment and detailed direction to support its expansion and new product market strategy.

With its acquisitions, LiftCo had gained volume. Nominal truck build capacity was increased to 7000 units per annum and employment to 3000. The company had also gained valuable export contracts in Africa and the Middle East, improving its home – export sales ratio, although these contracts were lost with political developments at the end of the 1970s. But LiftCo remained relatively small for a volume producer and, in comparison with other firms pursuing a mass production strategy, investment in new technology and improved facilities was limited. Clark Equipment, for example, had invested £15 million at its assembly site in West Germany to treble output to 12,000 units and to give the firm a world-wide capacity of 45,000 units per annum. Hyster's new plant in Northern Ireland, designed for the mass-production of electrically-powered vehicles in the low-lifting capacity range, cost £25 million to build and equip with robotics for use in the fabrication of subassemblies. This plant was strategically located, with the firm able to subsidize its investments through government development grants (*Materials Handling News*, July 1982:4). Throughout the 1980s, this multinational has been able to persuade the UK government to extend further financial assitance to ensure its continuing investments in other politically-sensitive high unemployment areas (*Guardian*, 21 September 1983).

Notwithstanding the merger activity of the mid-to-late 1970s, capital ownership in the British industry remained relatively fragmented at the end of the decade. New capacity was coming on-stream in the UK and in the international fork lift truck industry in the context of existing overca-pacity, and demand conditions were deteriorating. In Britain, political conditions were changing. Government policy had shifted towards a mone-tarist preoccupation with public expenditure cuts. LiftCo had acquired the second of its subsidiaries in 1977, or shortly before the appointment of Michael Edwardes to head British Leyland's board of management. This appointment confirmed the disintegration of the Labour government's commitment to modernization; from investment and technical change, an immediate reduction in British Leyland's losses was to be secured through the closure of capacity, or disinvestment. These developments in turn signalled the political ascendancy in Britain of neo-liberal perspectives on restructuring, with industrial change now to be enforced through the strengthening of market forces.

Market-led Restructuring
The Conservative government's support for high interest and exchange rates from 1979 had an immediate impact on profitability for firms in the UK lift truck industry. Export sales were adversely affected, while the collapse of UK manufacturing threw the domestic market into recession. At the same time, Japanese import penetration of the depressed UK market was rising. With the loss of its recently-acquired export contracts, LiftCo had been forced into a heavier reliance upon the domestic market. While there had been limited investment in the company, at the end of the 1970s management were still in the process of planning the changes required to support the firm's 'expansionary strategy', and had barely embarked upon the rationalization of product lines and reorganization of production facilities that had been intended to succeed its growth through acquisitions. The firm retained a complex model range, and triplicated

machining, fabrication and assembly facilities. In addition, it had acquired a top-heavy and complex management structure, weighing as an overhead and constraining productivity.

With deteriorating market conditions and intense competition from Japan, LiftCo initiated a destructive round of price-cutting in an attempt to maintain output. Since most competitors responded likewise, this did little to improve the firm's financial situation. The company registered a loss of £14 million on a turnover of £45 million in 1981. Firms in the Special Products Division had already been earmarked for privatization as part of the process of 'slimming down' British Leyland. Finding a buyer for LiftCo, however, proved to be no easy matter.

In an attempt to maintain profitability, other British companies and US-owned subsidiaries operating in Britain had reduced employment.[8] Nevertheless, and in spite of squeezed profit margins and intense market competition, firms displayed a 'remarkable resilience' to the pressures for rationalization (*Financial Times*, 23 October 1986). Commenting on the need for further amalgamations in the British-owned sector, one report on the industry in 1980 noted that the 'initiatives for improvement [lay] with the companies themselves', given the non-interventionist stance of the Conservative government, and the fact that the Sector Working Party appeared to have 'run out of steam', achieving little since the mergers of the mid-to-late 1970s (*Financial Times*, 2 May 1980). With two major privately-owned firms left in the British industry, neither of which appeared to be willing to withdraw, the Industrial Trucks Sector Working Party could achieve little and was eventually axed. Neither Lansing Bagnall nor Lancer Boss, moreover, appeared to be anxious to acquire LiftCo when it was placed 'on the market' in 1980. By this stage, the company was hardly an attractive, profitable asset, and competitors may have anticipated that it would simply be shut down. The lifting of exchange controls in 1979 had also enhanced the attractions of overseas investment for British firms. Thus, Lansing Bagnall indicated that the acquisition of LiftCo would not give it any 'significant volume advantages', that there would be 'considerable overlap of products', and that if that firm could afford to expand, 'it would be better to add to its overseas interests' (*Financial Times*, 2 May 1980).

Since LiftCo had proved difficult to sell, and government policies ruled out support for its development, it fell to British Leyland to finance the closure of capacity and redundancies. Escalating unemployment in the external labour market muted union resistance to job losses and inhibited the development of a united union stance to plant closures across LiftCo's subsidiaries. The company's recently acquired sites were shut, and production concentrated on the West Midlands' facility. Employment had been reduced to 1500 by the end of 1981, when it was announced that LiftCo had been acquired, indirectly, by the firm's major UK-owned competitor. The owners of this firm had established a new holding company in order to operate LiftCo as a separate enterprise, or one reliant upon the internal generation of finance for fresh investment. LiftCo's 'sale' to a UK-owned firm which had formerly declined enquiries from British Leyland had been prompted by reports that the company would otherwise be acquired by one of the Japanese lift truck corporations.

The UK within the International Lift Truck Industry
Lansing Bagnall and Lancer Boss have continued to expand through overseas acquisitions since the early 1980s, and have acted also to develop their UK-based facilities. The activities of these firms have contributed to radical changes in the structure of the international fork lift truck industry, changes which have been in progress since the 1970s, and which have been accelerated by economic crisis and corporate attempts to restore profitability. Thus, while in 1980 20 firms controlled approximately 60 per cent of world output, employers have since been predicting that, by the 1990s, 70 per cent of world production will be under the control of just 10 companies (*Materials Handling News*, February 1982:10). The battle for survival, and for domination within the industry, has been waged between British, West European, US and Japanese companies.

In Western Europe, firms have acquired their weaker rivals in an attempt to eliminate competition and to advance their position in world markets. Employers have also collaborated in the pursuit of protectionist policies to restrict Japanese import penetration, while individual companies have invested to improve their domestic facilities. Employing advanced technology in the manufacture of fork lift trucks, West European and, in particular, West German companies, have also shadowed Japanese firms in the development and production of automated materials handling systems. In the United States, in contrast, a number of companies have responded to Japanese competition by relocating the production of subassemblies to low-wage economies, with the internationalized character of the US industry inhibiting employer collusion in protectionism (*Financial Times*, 29 July 1987). Nevertheless, two US multinationals continue to invest in Britain.

The process of restructuring is as yet incomplete, and with an estimated 50 per cent overcapacity throughout Western Europe and the UK, further corporate failures are predicted (*Financial Times*, 29 July 1987). Employment in the industry in the UK therefore remains vulnerable. Inward direct foreign investment in this country has proceeded over the 1980s with European firms now expanding into Britain. Hence, the industry in the UK continues to be distinguished by the high representation of foreign-owned subsidiaries, with implications for the level and composition of employment and for industry performance in the future. Foreign-owned subsidiaries have to date largely supplied the relatively restricted demand for new materials handling technologies in the UK, with these firms locating their research and development facilities outside this country. Reports analysing

Britain's relative backwardness in the manufacture and application of the new technology have contrasted the distinctively abstentionist stance of the British government with the state support extended to the materials handling industry in competitor economies, in the form of subsidy, training programmes and government purchasing contracts favouring firms adopting new handling technologies (NEDO, 1983; 1986). Thus, the NEDO Mechanical Handling Economic Development Committee – which remains an active pressure group for change, although one with limited powers – has stressed the need for a strategic approach, at a national level, co-ordinating the mechanical handling, electronics and end-user industries, 'not only to assist the growth of the UK mechanical handling industry but also to help maximise the efficiency of UK user industries' (NEDO, 1983:7).

Managing to Survive

LiftCo was returned to the private sector at the end of 1981. Plant closures had reduced the firm's losses but it was still in a precariously weak financial condition with, at the time of privatization, industry commentators giving the company little more than a 50/50 chance of survival. Rigid financial controls were exercised by the new holding company, to which management responded with a series of initiatives over the period 1982–5. These included, in the first instance, further employment cuts and the scrapping of two-thirds of machining capacity. Aggregate employment had been reduced to under 1000 by 1985. In an attempt to restore and enhance profitability, however, management embarked on a more comprehensive package of changes, involving the reorientation of the firm's product market strategy and encapsulated under the epithet of 'bespoke engineering'.

Bespoke Engineering
Over the period from the late 1970s, LiftCo had been transformed from a medium-sized company to a small firm in an industry increasingly dominated by huge conglomerate corporations. Nominal truck build capacity had been reduced to around 2000 units per annum. But with overcapacity in the industry, and facing intense price competition from US and Japanese companies in the volume sector of the market, the firm was hard pressed to operate at this level of output.

In an attempt to expand output and sales revenue, management determined that the firm would concentrate on bespoke engineering, or the manufacture of 'tailor-made trucks for specialist lifting applications' and, over the period 1982–5, new models were launched with the aim of capturing specialized and less price-competitive niches of the market. In order to maintain capacity utilization, however, the firm would continue to compete in the volume market. By advertising sales features other than price – including product quality and the firm's ability to adapt basic designs to end-user specification – management hoped to differentiate LiftCo's products from cheaper, competitor models in this market sector. But the principal requirement was to reduce unit costs, in order to retain and increase market share in this product line, and to raise profit margins on other models in the range. Thus, while bespoke engineering apparently signalled a retreat into the past, in terms of the firm's marketing policy, it embraced changes in product design and in the organization of production designed to cut costs and to restore and enhance profitability.

The new system of 'module production', introduced incrementally over the period 1982–5, represented an adaptation of the JIT production and inventory control system employed in the Japanese fork lift truck industry and through the wider automotive sector in Japan (Abegglen and Stalk, 1985: ch. 5). The principle underlying this method of manufacturing is that component parts and subassemblies are produced for immediate incorporation into the final product, rather than for stock. It is generally contrasted with the Fordist flow-line mass assembly production system which involves the use of special-purpose machinery and production in large, homogeneous batches, in order to minimize the 'unproductive' labour and machine capacity involved in retooling product lines, and under which firms often carry large 'buffer' stocks to guard against the occurrence of faulty parts (and/or disputes at supplier companies) disrupting the continuity of production at the assembly stage. Representing 'idle' capital, inventories of work-in-progress and buffer stocks of parts are an expensive overhead. The progressive reduction of inventories serves to accelerate the circulation of capital from the intake of raw materials to the despatch and sale of final products and, with this, raises capacity utilization within the plant. Productivity increases are therefore achieved through speed-up and through work reorganization, with a distinctive set of working practices normally associated with the JIT production system (Sayer, 1986).

Inventory reduction is achieved through a variety of means, but principally entails changes in plant layout, in the arrangement (and/or type) of hardware at preassembly stages of production, in production control procedures, and in the firm's relationship with external component supplier companies (Schonberger, 1983). For example, the rearrangement of hardware (and/or use of programmable technology) at preassembly stages reduces machine set-up times. Partially offsetting the economies of long production runs (under the Fordist system), reduced set-up times (quick machine changeovers) allow firms to balance more accurately the rate of output of the various processing steps involved in production, so that parts are produced for immediate use at the next processing stage, and so on. Quick machine changeovers can therefore make it economical for firms to produce in short batches and to manufacture and assemble a more varied model range, without jettisoning the benefits of standardization. It is this potential of JIT that has merited its classification as a form of flexible manufacturing, enabling firms formerly engaged in the mass production of a limited product range to develop a greater responsiveness to market openings (Abegglen and Stalk, 1985: ch 5). Rearrangements facilitating a reduction in optimal batch size, on the other hand, have been interpreted in the 'flexible specialization' thesis as the means through which small firms formerly unable to attain scale economies can now aspire to achieve a more competitive footing with large-scale producers (e.g. Piore and Sabel, 1984; Sabel and Zeitlin, 1985).

While the system has been most extensively adopted by firms and in industries engaged in mass-assembly manufacturing, elements of JIT production have a far wider scope of application (Sayer, 1986:57), and it has been argued that JIT principles can be extended to stream-line job shop production processes (Schonberger, 1983:15). Nevertheless, in the international fork lift truck industry, economies of scale remain – and indeed, have increasingly become – significant in terms of the nature of competition. Thus, and following the Japanese lead, West German companies have standardized designs to secure scale economies in the manufacture of subassemblies, and in this way have been able to achieve economies within the manufacture of a more diversified model range. Thus, subassemblies are designed to be interchangeable across a range of models, or for subsequent assembly in mixed sequence batches under this particular variant of JIT module production (*Financial Times*, 29 July 1987). Larger firms have also been to the fore in this industry in terms of investment in labour-displacing advanced technologies at preassembly stages of production. LiftCo, in contrast, was financially constrained from pursuing such investment in the early 1980s; the company operated with an almost perpetual cash-flow crisis.

Moreover, the effective operation of JIT – or the ability to operate with minimum inventories – demands both a highly disciplined (or 'co-operative') workforce and a high degree of standardization in production schedules (in the level of output and sequence in which models are assembled). The latter is essential in order to maintain capacity utilization at preassembly stages and to support JIT parts deliveries, with supplier firms expected to synchronize parts deliveries to meet the buyer company's requirements (Sayer, 1986:55). The ability to predetermine output schedules in turn implies that market power and firm size remain important competitive assets.

At LiftCo, management's decision to pursue bespoke engineering reflected the firm's altered and now marginal status within the fork lift truck industry. Thus, management conceded that the firm was ill-equipped to 'compete head-on' with larger and more capital-intensive US, Japanese – and West German – producers, and would therefore have to nurture more sheltered market niches and attempt to secure sales orders by emphasizing the firm's ability to adapt basic designs to customer requirements. Operating with a short order-book, however, and increasingly forced to accept orders for 'one-offs' or prototypes in an attempt to maintain output and sales revenue, management were unable to restructure the firm's relations with external suppliers and with this secure improvements in productivity and cash-flow through the introduction to module production.

Module Production

The introduction of module production at LiftCo in the period 1982–5 was aimed, in the first instance, at resolving bottlenecks in vehicle assembly arising with the recurrence of parts shortages. Difficulties with materials supply were acute in this period, and were in part the product of management's attempts simultaneously to reduce inventory levels and extend the product range. Inventory levels had been progressively reduced since the beginning of the 1980s in the drive to cut overheads and to improve cash-flow. Management were attempting to circumvent the cash-flow problem by using revenue raised from sales to pay for the bought-in items required to complete these sales orders. But accepting orders for customized vehicles and 'one-offs' in order to maintain output, and operating on a short order-book, militated against the scheduling of input parts requirements. As a result, parts required to complete actual sales orders were often unavailable, with semi-built trucks stocked in the rectification area awaiting parts deliveries. The bulk of monthly output was being produced in the last two weeks of the month – when management could be assured of receiving payment for sales orders – and completed only with the assistance of high levels of overtime working. At the end of the month, fitters were often recustomizing the same vehicle two or three times; taking parts from a finished truck to complete another for which a more immediate payment had been promised, and subsequently rebuilding the 'donor' model.

As it involved changes designed to reduce vehicle build times and to secure a more perfect fit between incoming sales order receipts and parts requirements, module production was intended to resolve these difficulties; to ensure the maximum (productive) utilization of labour and machine capacity throughout the month; and with this, to secure an improved cash-flow position. Key elements in the move towards module production were technical innovation in the design and production control functions; changes in product design, in plant layout and truck build sequence; and work reorganization.

The company had invested in a CAD system, an investment which had, in the short term at least, exacerbated its cash-flow crisis. Reducing the labour time involved at the R&D stage of production, management hoped that the new technology would give LiftCo the flexibility to respond to the volume and variety of incoming sales orders for customized vehicles, whilst continuing to engage in new product development. The new technology was employed in the redesign of existing models in the firm's range and in the development of specialized vehicles. Models in the firm's volume range, for example, were modified to reduce the labour time involved in the production of each unit by an estimated 50 per cent. Design changes permitted new materials to be employed in fabrication, eliminated various stages in the production process, and meant that subassemblies could be slotted together at the final assembly stage, thereby eliminating the need for overhead cranage. Models were also designed with the aim of standardizing as many subassemblies as possible across the product range, so that the same parts could be assembled to complete different final products. In addition, new models were designed to be manufactured in-house, using the firm's own fabrication facilities to produce a higher percentage of subassemblies. This would raise capacity utilization in the fabrication department, and profitability through the manufacture of a higher value-added product.

Design changes on the trucks in turn facilitated changes in the layout of production and in build sequence, intended to reduce production lead-times and with this to raise capacity and labour utilization. This involved the relocation of subassembly work from the truck assembly plant to the fabrication department, with the fabrication and subassembly stages of production brought together to approximate more closely a continuous flow-line. Subassemblies, including the masts, overhead guards, chassis, axles and a range of special attachments, would be built up as a series of modules – or discrete product lines – in the fabrication department. The overhead guards, for example, would be fabricated, painted and assembled in this area in a streamlined process. The modules would then be transported across the town to the assembly plant, where the tracks were to be replaced by a series of stalls. Bought-in items, including the powertrain and ballasts, would be delivered straight to the stalls, along with a kit of smaller parts designated for a specific customer order. A two-man gang allocated to each stall would be responsible for assemblying the modules and other parts to complete the end product, following written instructions to attach any particular customized feature.

New working practices, introduced in the period 1982–5, were intended to intensify work and to contribute to the goal of fast throughput and low inventory production. Aside from the closure of machining capacity, job losses had been weighted towards indirect hourly-paid and clerical employment areas. The task of quality inspection, formerly performed by separate ancillary grades, had been reallocated to direct production workers in fabrication and assembly, with these workers now expected to assume 'personal responsibility' for the quality of their work. This was part of management's new quality control procedures, the aim being to detect and trace faults back to their source and prevent their recurrence. Aside from being a 'good advertising mark', the new procedures offered potential economies in parts purchasing and end-user warranty claims, and were intended to obviate the possibility of faulty parts bringing production to a halt with the move towards stockless production. They had also enabled management to reduce the quality inspection department by 50 per cent, the workers remaining in employment upgraded to the fitters' rate as part of the drive to achieve 'full flexibility' between assembly, fitting and test work. With full flexibility, management could reallocate labour within the plant in accordance with current production schedules and hence ensure maximum labour utilization.

Since direct workers now had to 'sign off' on the completion of each task, they were subject to more intensive managerial surveillance, while the changes in vehicle design and in plant layout had further diminished the skilled, or creative, elements remaining in the work of fitters in vehicle assembly. Management's ability to monitor progress in production, moreover, was enhanced with the changes in production control introduced as part of the move towards JIT manufacturing. Management hoped to achieve the goal of fast throughput, low inventory production with the assistance of a computerized production and inventory control (or materials requirement planning – MRP) system. With the new technology, data processing economies were secured in the parts purchasing, materials control and production scheduling functions, facilitating the reduction of clerical employment in these areas. The MRP system centralized and enhanced management's control over, and ability to monitor the performance of these different departments and, linked-up with CAD, secured a closer integration of their work with that of the engineering and sales functions. The aim, with the new technology, was to secure a more perfect match between incoming and anticipated sales order receipts and input parts requirements; to schedule parts deliveries (and initiate in-house fabrication) for immediate use in assembly, and the build programme for immediate sale and despatch. In principle, this should have eliminated bottlenecks in vehicle assembly and, with this, acted to intensify work through the speed-up of the production process.

In practice, however, bottlenecks continued to constrain production throughout the period 1985–6 and, indeed, the difficulties in manufacturing had become more acute. Thus, while the use of MRP had reduced the time involved in scheduling parts supplies, it had done little to enhance the firm's ability to ensure that the correct parts were delivered by external suppliers at the correct time to proceed with a truck build programme that would match incoming sales order receipts. In spite of moves to in-source a greater percentage of fabricated subassemblies, the firm's requirement was still for a diverse range of bought-out items. Indeed, with bespoke engineering, the range of different parts required had been increased, and the batch size of orders placed with suppliers reduced. This had combined with the firm's adverse cash-flow and current market conditions to undermine management's ability to restructure relations with external suppliers along the lines demanded by JIT production.

Difficulties with materials availability had been a feature of production at LiftCo in the 1970s, when the firm carried high levels of inventories and had a full order-book. These difficulties emanated partly from the firm's status with component suppliers. As the Industrial Trucks Sector Working Party had noted in the mid-1970s, firms in the fork lift truck industry are not priority customers for many of their suppliers whose fortunes are tied to the much larger motor vehicle companies within the automotive industry (NEDO, 1978). LiftCo's position in relation to its larger suppliers had been further weakened by the reduced scale of production in the 1980s, reducing its aggregate parts requirements, and by bespoke engineering which meant the purchase of a diverse range of parts in small batches. In the context of depressed and highly competitive product markets, moreover, the firm had increasingly been forced to accept an irregular sales order intake, which meant that it was making an irregular demand on suppliers for parts. LiftCo did not have the weight to demand JIT deliveries, while at the same time suppliers were being exacted to fulfil such an arrangement by larger corporations in the automobile industry. On top of this was the firm's cash-flow problem which management were attempting to resolve with module production. LiftCo had a poor reputation for paying for parts deliveries promptly. While the firm could exhort its smaller and more dependent subcontractors to accept deferred payment, other firms retaliated by demanding 'cash on delivery'. With these difficulties, therefore, bottlenecks continued to constrain production which in turn emphasized management's dependence upon labour's co-operation, with high levels of overtime working required to meet output targets.

Workplace Labour Relations
Analyses of JIT have emphasized that the effective operation of this production system is dependent upon the firm's ability to restructure relations with external component suppliers and internal labour-management relations (Sayer, 1986; Dohse *et al.*, 1985). The system is normally associated with a distinctive set of working practices designed to achieve flexibility in the utilization of labour. As noted elsewhere, however, productivity increases are also dependent upon the firm's ability to create a highly disciplined workforce, motivated to respond 'flexibly' to the requirements of the production system. Operating with reduced inventories and buffer stocks, in other words, means that workers must be disciplined '"to do on the spot" whatever is required' to maintain the continuity of defect-free production (Schonberger, 1983:14). Operating with reduced inventories, moreover, means that the continuity of production may be easily disrupted within the plant or at the firm's external suppliers. Hence it has been noted that, as operated in Japan, JIT is normally surrounded and supported by a wider set of labour management practices (and conditions in the external labour market) which serve to buttress managerial authority – and to minimize union influence – within the manufacturing plant (Dohse *et al.*, 1985). Opinion is divided, however, as to whether these employment practices can be successfully transferred to the West.

Dohse *et al.*, for example, have argued that the superior productivity of the 'Japanese management system' is possible only in 'an industrial relations environment in which there are hardly any limits to management prerogatives' (1985:140), and that such an environment is in turn unique to Japan, and follows the historically-specific development of capital–labour relations in that country.

Mass unemployment in the UK since the late 1970s, however, coupled with the enactment of restrictive employment legislation, has strengthened management's ability to introduce new forms of work organization and new employment practices. At LiftCo, for example, unemployment in the external labour market and internal job losses set the context in which the workplace union organization had acquiesced to the introduction of 'flexible' work arrangements and new payment systems in the early 1980s, in the hope of protecting the jobs of workers remaining in employment. In the period of the redundancies, moreover, the cohesion of the workplace union organization had largely been torn apart, with different areas and grades placed in competition for jobs, although solidarity was to an extent rebuilt by management's attempts to by-pass the union and established collective bargaining procedures. Under pressure from the banks and the firm's new owners to execute cost-cutting measures and to restore profitability, management had attempted to introduce new working practices unilaterally.

Management at LiftCo attempted to assert the managerial prerogative of the 'right to manage', in the face of shopfloor opposition to breaches of procedure. But given the constraints besetting production, management were dependent upon labour's co-operation – particularly in the area of overtime working – to maintain output. Furthermore, and in spite of the use of micro-electronics technology in the design and production control functions, management remained dependent upon labour's creativity and ingenuity in resolving the difficulties arising with parts shortages, and in dealing with the engineering department's overload of customized sales orders. Thus, fitters and chargehands in assembly were frequently presented with sales orders for which engineering designs had as yet to be processed. The necessity to work out how to adapt such vehicles – and how to find the necessary parts to build them – in the context of the 'end of the month rush' to get trucks out was experienced on the shopfloor as work intensification. Nevertheless, workers expressed their recognition of the company's difficulties, and their obvious interest in company survival, and hence extended their co-operation to complete the build programme. Co-operation with overtime working was also induced by worker interest in maintaining earnings, and since the district union office perceived that this was the only way to keep the firm in production, the national agreement on overtime working was regularly suspended at management's request. The overtime ban, or at least the threat of the withdrawal of co-operation, however, had become a sanction that could be employed with effect against management's attempts to by-pass established procedure. Labour's ability to employ this purely defensive sanction was in turn the product of the contradictory pressures and relationships within which the company was operating and which management were unable to resolve autonomously.

Conclusion

This chapter has examined the unfinished process of capital and labour restructuring in the fork lift truck industry. It has focused on the specific circumstances of one British company and explored the reasons for its relative decline within the UK industry since the mid-1970s. The analysis can be reviewed in relation to current debates about 'flexible specialization'.

The ability to combine the benefits of mass-production, in terms of unit costs, with product diversification, clearly presents attractive opportunities for individual firms searching for new market openings, or seeking to reduce unit costs within an established arrangement of short-batch production. In recent analyses, however, the use of microelectronic technology, 'flexible' forms of work organization, and new parts supply arrangements in the short-batch production of quality, customized items for distribution to specialized market niches, has been identified as a new model of competitive behaviour to which firms, regions and countries must adjust (see, e.g., Best, 1986). Central to this perspective on contemporary capitalist restructuring is the proposition that changing patterns of consumption have undermined the profitability of enterprise organized for the mass-production of standardized goods. New patterns of demand and the increasing sophistication of consumer tastes are argued to have opened up new, profitable market opportunities which, with the assistance of advanced flexible production systems, firms can now exploit and enjoy a return to prosperity through the pursuit of economies of scope rather than scale.

The new synthesis of technology and markets is viewed as a basis for a generalized economic recovery (Piore and Sabel, 1984). It is also argued to be the basis for a transformation in industrial structures and in capital–labour relations at the point of production. Considered to be more innovative – or responsive to rapid shifts in micro-markets – small firms are granted something of a competitive edge over large, bureaucratically-organized corporations in the new regime of 'flexible specialization'. The new forms of work organization supporting flexibility in production are said to enhance the intrinsic interest of work for production employees, in comparison with the detailed division of labour associated with mass production.

The 'flexible specialization' thesis has been criticized for its preoccupation with developments in the sphere of consumption, and its interpretation of current conditions in product markets (Gough, 1986:63). The analysis, moreover, pays scant attention to the specific circumstances of capital goods sectors, although product and process innovation in such industries are recognized to be vital components in a generalized economic recovery. In the post-war period in many of these sectors, including the fork lift truck industry, mass production and the pursuit of scale economies were of limited significance in terms of the nature of competition. In the international lift truck industry, as in other investment goods sectors, this position has been reversed more recently (Gough, 1986:64). Under the lead of Japanese competition, firms have restructured production to secure a reduction in unit costs in the manufacture of components and sub-assemblies through standardization in final product design, the introduction of labour-displacing advanced technology, work reorganization and new inventory control systems. While Japanese producers have focused production for the volume sector of export markets, however, European companies have organized production so that common modules can be employed in the assembly of different final products. These competitive responses, nevertheless, have so far failed to alleviate the pressures on profitability exerted by surplus productive capital in the industry. Hence, while there have been fundamental changes in industry structure, competition remains intense and further corporate failures are predicted.

Underwriting the trend towards the increasing concentration of production in the international fork lift truck industry is the necessity to engage in major product innovation in order to create and exploit new, profitable market opportunities. The move into the manufacture of automated handling systems has raised entry costs in this sector because of the need to devote resources to R&D, and to organize production for the manufacture of systems technology rather than discrete items of mechanical handling hardware. Large, multi-plant firms have thus dominated the production of automated materials handling equipment. While sheer size has been advantageous for firms seeking to establish a leading position in world markets for the new technology, however, it has not been sufficient to guarantee the profitability of product innovation. The industry's ability to reinvigorate aggregate demand is restricted by its dependence upon the development of end-user consumption and service goods sectors. Governments in most advanced economies have acted to support the development of the new technology – through subsidy, 'user-awareness' programmes and favoured purchasing agreements – with an increased scale in the manufacture of AHS serving to reduce the cost of the new technology relative to labour costs under established technologies.

Reports have drawn attention to the relatively limited support extended by the UK government to the materials handling industry, and related this to the relatively slow adoption and manufacture of AHS technology in the UK in the period to the mid-1980s (NEDO, 1983; 1987). With respect to the fork lift truck subsector, however, the present analysis has argued that the adjustment of the industry in the UK to the patterns of change and restructuring in the international industry has been shaped by state policy in interaction with the distinctive economic and social structures of this sector in Britain. These include the structure and composition of capital ownership, with the UK distinguished by the high representation of both foreign-owned subsidiaries and of relatively small-scale British producers. Thus, the diversity of capital organization within the industry has inhibited the development of a united employer approach towards change and industry development in Britain.

State policy in relation to this sector of UK engineering has shifted over the period from the 1970s from attempts to promote voluntary restructuring via indicative planning to a more overt reliance upon market forces. Whereas the trade unions had been party to the planning initiatives of the 1970s, their ability to act as a force for industrial change was undermined, in part, by fragmented labour organization. But indicative planning proved a weak mechanism through which to direct and co-ordinate the development of the industry. Since 1979, government policy has supported British firms in their search for profitable investment opportunities abroad, while labour market conditions have facilitated the introduction of new working practices in UK-based operations. Nevertheless, surplus productive capacity continues to constrain productivity growth in the UK industry. Government exhortation for wage restraint, moreover, conflicts with the stated objectives of productivity growth and with the reorganization of end-user sectors on the basis of advanced materials handling technologies.[9]

As a subsidiary of a state-owned corporation, LiftCo has been vulnerable to the shifts and inconsistencies in government economic and industrial policies. It has been argued that the attempts to enhance productivity under British Leyland in the 1970s through the reorganization of production to secure manufacturing scale economies were undermined both by the relative underinvestment in the company, and by the absence of a coherent strategy for the development of the British fork lift truck industry. Support for the firm's expansion was finally terminated with the political transition to market-led restructuring, which precipitated the collapse of demand at the end of the 1970s, and forced British Leyland to accelerate the process of retrenchment.

The pursuit of bespoke engineering – or the manufacture of quality, customized trucks and development of specialized market niches – following privatization reflected the firm's altered status within the industry and weak position in relation to large-scale competitors, and was not a response to changing patterns of demand. The firm's chronic cash-flow position, and inability to secure finance via the new holding company, forced management to rely on work-intensifying forms of production reorganization in their attempts to raise productivity. Management's ability to restructure production and secure the effective operation of JIT, however, was in turn constrained by the firm's weak position in overcapitalized product markets. The difficulties in production, and reliance upon overtime working to complete monthly truck build programmes, enabled the workplace union to maintain its influence within the company. Labour's ability to influence the wider set of relationships constraining the restoration of profitability through workplace union activity alone, however, was minimal.

LiftCo was forced into liquidation in 1986, and the firm has subsequently been acquired by a European-owned multinational competitor company. The change of ownership has brought fresh investment and the direction required to support the development of the West Midlands facility. Nevertheless, with the persistence of overcapacity in the international fork lift truck industry, employment in the UK sector remains vulnerable with firms engaged in a battle to enhance their position at the expense of competitors.

Notes

1. This chapter is based on interviews with management, shop stewards and shopfloor workers at the case study company over the period 1983–5. The interview data are supplemented by information from company reports and shop steward records, financial and trade press articles, and government statistical publications.
2. ICC Business Ratios (1977) cited the mechanical handling industry (MLH 337) – which embraces industrial trucks, conveying, lifting and winding, lifts and escalators – as one of eight sectors with an average return on capital of over 20 per cent in the mid-1970s.
3. One of the US multinationals – Clark Equipment – relocated from the UK in the early 1970s to concentrate its European manufacturing operations in West Germany.
4. UK sales of fork lift trucks registered between 15,000–20,000 in the 1970s, which compared to a US domestic market of 80–100,000 units.
5. Balkancar, the world's largest lift truck producer, which serves the Commecon market, was engaged in the mass production of lift trucks throughout the post-war period.
6. Notably, a number of the foreign-owned subsidiaries operating in Britain.
7. Less strategically, other US companies simply added a new 'low cost' line to their existing product range (*Materials Handling News*, November 1982).
8. Employment data on the fork lift truck industry are subsumed within the wider materials handling industry in government publications. Between 1977 and 1982, employment in this larger grouping was reduced from 60,300 to 47,500 (*Department of Employment Gazette*).
9. Certain parallels can be drawn with the immediate post-war period when Britain was reported to be lagging some 10 years behind the US in terms of the manufacture and application of mechanical handling equipment. Melman (1956: 167–8) noted a conflict between government appeals for wage restraint and government identification of productivity growth as a 'highly desirable objec-

tive' in the UK in this period. He concluded that the narrowing of the productivity gap between the US and Britain achieved over the 1940s and 1950s was due in practice to the 'successful wage pressure of British workers', mounted in spite of government policy emphasizing wage restraint, and forcing employers to invest in labour-saving, productivity-enhancing technologies.

References

Abegglen, J.C. and G. Stalk Jr. 1985. *Kaisha: The Japanese Corporation*. New York: Basic Books.

Best, M. 1986. 'Strategic Planning and Industrial Policy,' *Local Economy*, No. 1, 65–77.

Business & Market Research. 1981. 'The Market for Fork Lift Trucks.' Unpublished report.

Dohse, K., Jürgens, U. and Malsch, T. 1985. 'From "Fordism" to "Toyotism"? The Social Organization of the Labor Process in the Japanese Automobile Industry', *Politics and Society*. Vol. 14, No. 2, 115–46.

Financial Times. Various dates.

Gough, J. 1986. 'Industrial Policy and Socialist Strategy: Restructuring and the Unity of the Working Class', *Capital and Class*, No. 29, Summer, 58–81.

Guardian, 1982. 'Japanese Lifters Set to Dominate the World', 18 August.

——. 1983. 21 September.

Materials Handling News. Various dates.

Melman, S. 1956. *Dynamic Factors in Industrial Productivity*. Oxford: Blackwell.

National Economic Development Office (NEDO). 1978. 'Industrial Trucks Sector Working Party'. Unpublished report, November.

——. 1983. 'Prospects for the Mechanical Handling Industry'. Report from the Mechanical Handling Sector Working Party, July. London: HMSO.

——. 1986. 'Advanced Handling Systems – Exploiting the Opportunities'. Report from the Mechanical Handling Economic Development Committee, April. London: HMSO.

Piore, M. and C. Sabel. 1984. *The Second Industrial Divide*. New York: Basic Books.

Sabel, C. and J. Zeitlin. 1985. 'Historical Alternatives to Mass Production: Politics, Markets and Technology in Nineteenth Century Industrialization', *Politics and Society*. August, 131–76.

Sayer, A. 1986. 'New Developments in Manufacturing: The Just-In-Time System'. *Capital and Class*. No. 30, 43–73.

Schonberger, R.J. 1983. 'Japanese Manufacturing Techniques: Nine Hidden Lessons in Simplicity'. *Operations Management Review*. Spring, 13–18.

Trades Union Congress. 1977. The Trade Union Role in Industrial Policy. Report of a Conference of Associated Unions to discuss the trade union role in industrial policy. October.

Shopfloor Industrial Relations

The limits of plant level trades unionism in influencing change have featured in all the case studies. This chapter examines the apparent stability of plant level bargaining structures in this period, and criticizes those approaches that analyse them in isolation from developments in product markets.

This analysis of the three manufacturing plants challenges the view that large-scale unemployment and government policies have left workplace industrial relations essentially untouched by considering the changing power relationships beneath the surface of institutional stability.

The idea that 'core workers' have been insulated from change is rejected in this study. While unemployment and changes in the labour-market do not directly challenge the position of those workers surviving rationalization, they form an inescapable constraint on shopfloor organization. Hence attempts to assess the relative strength of managers and union organization can be highly misleading if only their formal aspects are considered. To understand fully the problems facing local organization they must be placed in the context of shifting pressures in the product market. On the one hand, the insecurity produced by market competition reduces union capacity for formulating and pressing their independent interests. On the other, it simultaneously determines the logic of managerial action. It is the interrelationship between these factors, rather than simply questions of union confidence, or management strategy, which defines the political problem facing labour.

In these circumstances, the limited reach of even the tightest shopfloor organization can render active resistance to management initiatives unlikely or ineffective, being unused to confronting problems of what is produced and how. At shopfloor level, therefore, the result may not merely produce acquiescence, but an active association with steps seen as maintaining the employer's market position. Such an association is itself far from new, and has already been seen as a significant factor at sectoral levels. What is new is the extent of the challenge that faces labour within work, and the location of the forces operating on it. 'Factory consciousness', in many ways reinforced by decentralized bargaining, can allow one site to be played off against another, while attempts at producing 'alternative' courses of development all too often show the enormity of the gap between union policies towards industrial democracy and planning, and the everyday experience of the membership.

7

Recontextualizing Shopfloor Industrial Relations: Some Case Study Evidence

Michael Terry

Attempts to understand the relationships between industrial restructuring and patterns of shopfloor industrial relations have been bedevilled by both failures adequately to conceptualize restructuring, on the one hand, and to contextualize shopfloor industrial relations, on the other. The former issue has already been dealt with at length in this volume; here I concentrate on the need to understand observed patterns of behaviour on the shopfloor in the context of the reorganization of production that has been taking place.

Most recent empirical accounts of workplace industrial relations in the 1980s have, like many that preceded them, concentrated on essentially procedural indices for establishing the extent of change. Such an approach, empirically convenient if not theoretically justifiable during periods of relative stability, loses much value during periods of fundamental change, since the theoretical inadequacy of separating industrial relations institutions from the organization of production is clearly revealed. One consequence of the use of such indices can be seen in the debate as to whether or not employers have been mounting a new offensive against shopfloor trade unions, in which the degree of change in a number of procedural measures is used as the basic evidence for or against such a 'managerial offensive'. Although this debate has thrown up much of interest, the reliance on procedural issues makes it ultimately unresolvable.

A second consequence, of greater significance for this chapter, has been for the content of related debates about the 'strength' of shopfloor trade union organization and the tactical response (or the lack of it) to the changes that have been taking place. On the former point, I have attempted to summarize the main points of the debate elsewhere (Terry, 1986). On the latter, one view is that shopfloor trade unions have been unaffected by the changes. There are two themes contained in this. The first is that at least among those workers left in full-time employment in manufacturing, unions are continuing to exert broadly the same influence on the introduction of new technology, work organization and job design as they have always done. This view is perhaps best exemplified by the work of Batstone and Gourlay (1986). The second variant, most recently stated by Daniel (1987), is that unions in this country have rarely sought to impede the introduction of change anyway, and in that sense, at least nothing has changed in the behaviour of shopfloor trade unions. These two views are compatible, even if they carry different shades of meaning. The opposing view, that might be summarized in the journalistic phrase 'new realism', is that shopfloor trade unions have been forced to come to accept the inevitability of unpalatable change. Among the 'new realists' two further distinct views may be discerned, one that holds that unions will regain their traditional position once the economic position turns to their advantage, and another that considers that some trade unions, at least, have perceived the short-sightedness of their previous policies of opposition and have fundamentally changed their approach.

The inconclusive nature of these debates again owes much to the unsatisfactory nature of many of the indices used, and of the decontextualized use of terms such as 'control' and 'influence' used to try to estimate the balance of power between management and unions. The central theme of this chapter will be that it is necessary to pay greater account than hitherto to the role of the product market in understanding both the dynamics of shopfloor industrial relations and the responses of shopfloor-based trade unions. A start has already been made by Hyman (Hyman, 1989). He argues that, in respect of full-time workers in manufacturing industry – 'primary' workers – the major force at work in securing the acceptance of change does not emerge directly from the state of the labour-market – the level of unemployment – but from the product market. Thus:

> The main effect of the recession on primary workers is *not* through changed labour market conditions; the reserve army of unemployed does not act as a competitive force undermining established conditions. Rather, it is changed *product* markets . . . which generate a (perceived) threat of job loss through closure or radical contraction of operations. In such circumstances, 'new realism' represents an acceptance – or even, in some cases, an active pursuit – of forms of production (re)organisation which sustains the employer's market position (Hyman, 1989:196; emphasis in original).

This characterization comes close to capturing the mood described by John Edmonds who has argued that

> As the recession deepened, the fear of redundancy touched every trade union family in Britain . . . Employees began to doubt the survival even of strong companies . . . Many managers have used the recession as an all-purpose reply to every union claim . . . extra productivity has been demanded without extra payment . . . Employees . . . resent the changes bitterly . . . but when . . . faced by management threats . . . are less willing to press their case (1984: 18–19).

The view of workers in the manufacturing sector as being (and perceiving themselves as) endangered rather than secure (more accurately, perhaps, significantly less secure than at any previous time in the post-war period) is a necessary corrective to the perception of 'core' workers as being insulated from the pressures of change. It also gives a starting point for understanding the relative acquiescence of workers and unions in the introduction of technological and organizational change observed in recent years (see, e.g., Edwards, 1987: 122–7). For it provides a context of worker and union uncertainty and consequent timidity within which to observe the processes surrounding the (negotiated) introduction of change. The state of continuous uncertainty about the future engendered in the workforce provides the backdrop against which both the introduction of managerial proposals for change and reactions to them must be assessed.

But even within this formulation the focus of attention is still restricted to a view that union power is primarily diminished by a lack of self-confidence, a fear of the consequences for workers' jobs of collective action or other forms of resistance to management. It does not fully recognize that the pressure of product markets does not only provide *negative* reasons for workers' acquiescence in the reorganization of production, it also provides a series of *positive* reasons why change must be introduced. The product market thus provides a basis for the *logic* of managerial action that has to be taken into account in our analysis. What I will therefore raise in this chapter is a second element in the relationship between product market change on the one hand and union structure and response on the other, namely that the very issues raised by the process, and the manner in which they are presented to the workforce, pose major problems for existing forms of workplace trade union organization. Workplace trade unions thus have to deal with what is both an institutional and an ideological crisis, as argued in the Introduction to this book, and it is meaningless to attempt to discuss and evaluate the one without the other.

It is necessary to recall that in one sense at least, there is nothing new in the observation that workers and their representatives have problems in dealing with arguments based in considerations of production. As Lane has recently noted

> Trade unionists . . . have not normally . . . concerned themselves with the logic of the enterprise. In the workplace this shows itself in an unconcern with the organisation of production unless it affects health and safety, the ability to earn, the division of labour . . . and the loss of jobs . . . it is *not* common for workers or their representatives to express concern about what is produced and how (Lane, 1986: 326, emphasis in original).

It is worth stressing that even where unions were concerned with the consequences of production organization, their agreed solutions often came back to pay ('dirt money', 'danger money' and so on) rather than to changes in the organization of production intended to remove the *cause* of grievance.

The particular strengths and weaknesses of this traditional lack of interest, and the corollary characteristics of economistic objectives and sectional organization have been fully explored elsewhere. Here the important point to make is that this tradition developed within a particular economic context, one which for many years was characterized by growth (or at least stability) on the one hand, and sluggish technological change and production innovation on the other. Within such a context the limited horizons of shopfloor trade unionism, concentrating on arguments about distribution and on the defence of the *status quo*, had a clear logic reflecting employers' own managerial priorities of stability and continuity. Union organization, perfectly logically, therefore crystallized around structures of distributive bargaining, while its politics accepted managerial production priorities and methods, and the 'right to manage' in these areas, provided at least that the money was there to pay for any occasional disruptions that might occur. Equally obviously, such logic no longer applies, with consequent problems both for the structure and the politics of trade unions in dealing with an employers' agenda set by crisis and change.

The need to raise these questions is reinforced by the unambiguous observation, made in other contributions to this volume and in the case studies below, that the changes are not neutral with respect to workers. Although the *problems* confronting industry in the 1980s may be increasingly identified with forces outside the factory (and this too is an important contrast with previous periods), the *solutions* frequently involve the dislocation or disruption of existing patterns of work. To that extent, workers may not any longer be a major part of the problem, as perceived by management, but they are a significant part of the solution. One would therefore expect their representative organizations – trade unions – to have an interest in shaping the outcome of change.

An important question thus emerges, namely the extent to which union structures and union ideology have been able to cope with the issues raised by the dramatically changed production environment of the 1980s. Such questions cannot be answered by looking at formal indices of procedure and structure, since it is perfectly possible that these may remain unaffected while what stewards and managers *do* within them has been radically changed. A similar substantive approach is also needed to provide a framework for the analysis of novel forms of employee communication and consultation (team briefings, quality circles, and the rest) the significance of which is only partly revealed by seeing them as a means of by-passing trade unions. It is the *message* they are carrying that is at least as important in understanding their contribution to shopfloor factory politics.

The purpose of the case studies detailed below is to present data on the consequences of product restructuring on shopfloor industrial relations and trade unions in three companies. They are not intended to provide comprehensive accounts, but to illustrate the changes that have been taking place outside the much-documented engineering sector. All are presented in a manner intended to cast light on the changes introduced into the organization of production, the means used and the arguments deployed, and the responses of the shopfloor trade unions. The cases have been selected in order to present views of four rather different aspects of the process. Thus company *A* details two plant cases, one facing total closure and the other, partly as a direct consequence, facing a period of relative market and employment stability. Company *B* deals with a plant still preoccupied with adjustment to decline and uncertainty that has lasted for several years, while company *C*, having been through just such a period was, by the time of the research, re-emerging as a growing, but very different, concern.

The data were collected from intensive interviews with personnel managers and shop stewards, over a period of at least three months during 1985–6, supplemented by analysis of records of managerial, union, and joint meetings, where available. Attention was focused on the manual workforce. In all cases the analytical focus of the research was the plant; information about company, national and international developments was only collected on a fragmentary basis, so there is little systematic information on developments at these levels, either for the companies or the trade unions.

It is not an objective of this chapter to argue that employer strategies for restructuring in response to product market change necessarily embody as an explicit central component the reorganization of the production process, still less that they necessarily have a manifest industrial relations component. Marketing and investment decisions, for example, may well be taken without any explicit consideration of labour relations factors. Nevertheless, it is argued that they will all have implications, more or less profound, for workers, and hence for their representative organizations.

The Case Studies: Some Background Evidence

Product Market Change and Job Insecurity

The plants looked at were all part of multiplant companies, one overseas-owned and two UK-owned. In all three companies, changes in market demand had led to substantial redundancy and closure as management reorganized to make more intensive use of capacity, to close down unprofitable lines and to change product mixes. At plant level, however, experiences differed; in one case total employment levels had remained at least stable since 1980, as other plants in the group were closed and production was concentrated on the remaining sites. In the other two plants employment levels of manual (and non-manual) workers had fallen sharply since the late 1970s, through a mixture of retirement and voluntary redundancy in both cases. All those workers who remained had, therefore, been recently exposed to the reality of the threat of job loss. In two of the plants, the shedding of jobs had been a virtually continuous process over a period of five or six years. In the same two plants, the financial – and hence employment – prospects for the short and medium term remained uncertain throughout the period of the research.

Stewards, and indeed most managers interviewed, perceived a clear link between falling demand for the products traditionally produced by the company, job loss, and worker behaviour. In the two factories where employment had contracted, uncertainty about future demand for the company's products was held to be the major reason for managements' continuing success in attracting volunteers for redundancy. The certainty of 'money now' appeared to be seen, even by younger workers, as preferable to continued employment for an uncertain period, with the possibility of redundancy on less favourable terms at some point in the future. And in the one factory where employment had remained stable, the knowledge that other plants in the group had been shut down in response to falling demand was seen as a powerful deterrent to militant action – or even to resistance of any kind to managerial proposals. One convenor described how the workers at one plant in the group had gone on strike against a company pay offer and had believed – wrongly – that management had been bluffing when they threatened sacking all the workers and closing the plant. Since that event in 1980, initial company pay offers had been accepted by workers in other plants in a ballot every year – twice against a strong steward call for rejection. The insecurity generated among workers by uncertainty about the future was certainly seen by stewards – and most personnel managers interviewed – as the feature which gave plant-level industrial relations in the 1980s their peculiar and novel features. Although the precise effects are impossible to quantify, there seems no reason not to attach considerable weight to this factor.

The Institutions of Plant-level Industrial Relations
In all three plants virtually 100 per cent of manual workers were in a trade union. Only in one case had the company stated its intention to withdraw from a Union Membership Agreement unless the unions held a ballot – which they refused to do. Even in this case, however, the company replaced the UMA with a formal commitment to recommend to all new employees that they should join the appropriate trade union. Otherwise some kind of post-entry closed shop persisted.

Nor had the structures of shop steward organization been affected. In all three plants, with manual workforces varying in size from 250 to 1000, there was at least one full-time manual shop steward. In one case (not the largest plant) there were two. However, of the four full-time stewards in the case studies, all except one were young – in their twenties or early thirties – and had held the convenorship for at most five years. These three had all replaced what they saw as convenors 'of the old school' – competent, but set in their ways. All had left the companies by taking retirement or on redundancy terms. Steward numbers had fallen more slowly than employment, with a consequent drop in constituency sizes. There was no evidence that steward turnover rates had changed, nor that there were greater problems than previously in finding volunteers. Stewards met as frequently, and enjoyed the same facilities as they had done in all cases for a decade and more. There was no evidence of any change in long-standing relationships with full-time union officials, and senior stewards continued to participate in union affairs through, for example, membership of national, regional, and sectoral committees.

Managerial support for plant-based union organization similarly remained unaffected: there had been no efforts to put full-time stewards back on their jobs, no tightening-up on rights to time off for union business, no new reported restrictions on steward mobility, or ability to represent members. Likewise the institutions of collective bargaining had largely remained unaltered. Wage negotiations – conducted in two companies at plant level by stewards, and in the other at company level by national union officials – continued ostensibly as before, as did the normal run of joint bodies to hear job evaluation claims, to distribute and regulate overtime, and to resolve individual grievances and disciplinary issues. Some sort of committee that could be described as having a consultative function was found in all the plants, but in only one could it be seen to be marking any clear innovation. In none of the plants had there been major changes in production technology, with the result that there had been no need to attempt to reach agreement on novel, and sometimes problematic, areas such as new technology.

In short, therefore, the plant-level institutions of industrial relations appeared to have survived intact the period of recession since 1978–9. This is in line with all the survey work done in recent years (Millward and Stevens, 1986; Batstone, 1984; Batstone and Gourlay, 1986). However, as

I will argue below, the impression of institutional stability and continuity created by this evidence is at odds with more detailed pictures of substantive change obtained from a study of the context and content of plant-level industrial relations in each of the plants studied. In order to illustrate this, findings from each of the three cases will be summarized below, as each has particular features of interest.

The Case Studies: Employer Adjustment and Union Response

Factory A

This factory was part of a large multiplant conglomerate in the food processing industry. Demand for its main, long-established products was in slow and apparently irreversible decline, having fallen by just over 4 per cent between 1972 and 1981. In addition, competition for scarce supermarket shelf space with a range of other products had intensified, as had the competition from 'own-label' brands of similar products. Efforts to retain market position against other products had been hampered by increases in import costs of several raw materials, most notably imported foodstuffs. There was little import penetration.

The company dominated the market in its major product line, accounting for 40 per cent of domestic market share by value. Its response to falling demand and uncertainty about the future was two-fold; it had rapidly diversified into a range of food-based industries through a series of takeovers and had sharply reduced its manufacturing capacity in its traditional product line by closing down one entire factory in the mid-1980s. The product diversification had few direct industrial relations consequences for the factory studied, except that it had prompted the company to attempt to develop a more formalized company-wide approach to industrial relations and personnel issues, which are discussed below. The factory closure, however, effectively had two quite distinct effects. In the factory designated for closure it led to frantic union activity both within and outside established bargaining and consultative procedures. For the remaining factories, including the one studied, it contributed to quiescence and stability. Since both episodes reveal something about problems faced by unions, they will both be covered briefly.

The Doomed Factory

Unions in the factory designated for closure responded in a manner that has become increasingly familiar (see Levie *et al.*, 1984); they produced an alternative 'workers' plan' identifying ways in which a slimmed-down workforce, higher productivity levels and more aggressive marketing, among other things, could turn the plant back to acceptable levels of profitability. To this end the stewards and union officials enlisted the expert assistance of both their own organizations and outside bodies. In collaboration with local churches and local authorities, they produced a 'social audit' testifying to the wider social costs of closure; and they used this document successfully to enlist the support of local church and community leaders and politicians. By contrast, the unions made few attempts to engage the support of workers and unions in other factories in the company, with whom they had traditionally had very few dealings. There had never been a combine committee, and contacts between stewards from different plants were restricted to annual meetings with full-time officials before company-wide pay negotiations. Multi-unionism and the different backgrounds of the once-independent family firms which made up the company exacerbated the problem.

But the plan suffered most from factors internal to the plant and the shop stewards' organization that sponsored it. First, in its origins and development it reflected the traditional exclusivity and hierarchy of the stewards' body. It was largely the preserve of the convenors and senior stewards who followed long-standing practice in failing to involve members and section stewards in'its development. One consequence was widespread membership uncertainty as to what was going on when the plan was unveiled at a mass meeting. A second problem arose over what to do with the plan once it had received membership endorsement. The initial reaction of the company had been to encourage the unions to produce their plan and to welcome their initiatives to keep the company open. However, according to the stewards at least, the attitude changed when it became clear that the 'workers' plan' appeared feasible within management's own terms; they became more hostile, and 'moved the goalposts' by increasing significantly their estimate of the required level of profitability to keep the plant open. At the same time the company effectively stopped talking with the trade unions. One consequence was that they were denied a forum in which to discuss their proposals with the company. All existing machinery for consultation and negotiation had traditionally been used to handle issues of terms and conditions, with unions responding to managerial proposals. The unions were stymied by a situation in which the rhetoric of consultation was dropped and where they, unlike management, were in no position to implement their proposals unilaterally.

Finally, the plan suffered from the fact that it contained a lot of bad news for the workforce. Although, if implemented, it might have kept the factory open, it would have done so only at the cost of many lost jobs, an intensification of work, and falling real incomes. By accepting the logic of the managerial case for closure, the unions were constrained to come up with a plan that would significantly improve productivity through increased worker exploitation. Small wonder perhaps that the workforce, largely kept in the dark through the plan's formulation, refused at another mass meeting to endorse steward recommendations for industrial action in its support. The 'social audit' element in the stewards' response, although containing the seeds of an alternative approach to production, was used solely to garner external political and community support. The plan put to the company followed strict managerialist – and capitalist – rationality.

Faced with what was in effect a rejection of their proposals by the company and the membership, the unions' campaign of principled opposition collapsed. Stewards fell back on to the routine practice of talking about money within parameters set by management and devoted their activities to negotiating the redundancy payments and the timing of the closure. The company, for its part, announced that no new recruits would be engaged anywhere until redundant workers had been given the opportunity to transfer. Since the nearest alternative site was over 60 miles away, few availed themselves of the offer.

The Survivors
At-corporate level, the firm had been restructured to take account of market changes, firstly by product diversification and secondly through a qualitative reduction in capacity. Industrial relations did not feature as a major issue in either the definition of the problem or in the solution. It is therefore not surprising to find little evidence of managerial attempts to reform the institutions of industrial relations. But this is not to say that the changes described did not affect the content or conduct of plant-level industrial relations.

The factory closure took out around 15 per cent of manufacturing capacity in a period when product demand was falling by less than 1 per cent a year. The company therefore concentrated on increasing output at the remaining plants, while constantly seeking to keep down production costs. Thus, in an effort to increase labour productivity, the company introduced two major 'de-manning exercises' in 1984 and 1986, which in effect took one worker out of each production group of 10 or more. Workers displaced by this exercise were to be transferred to new work coming into the factory from the closed plant. The union branch records show that its outright opposition to these proposals disappeared within a fortnight, to be replaced by a policy that the selection of workers to be moved would be based on factory service. Company agreement to protect earnings up to two years after transfer helped, but the crucial point in defusing any potential union resistance was that no workers lost their jobs in this factory because work was being transferred from the closed plant. Significant labour productivity increases (company figures show an increase of over 11 per cent over a four-year period) were obtained without workforce loss at plant level; their smooth implementation was facilitated by job loss at company level.

As existing production was concentrated on to fewer plants, two further issues were dealt with by management. One was the increasing production time lost as machine breakdown times increased (little investment in better machinery was contemplated in a declining market). The second was the need to introduce increased amounts of shiftwork to improve capacity utilization on lines with high product demand. The first problem was attacked in two ways. First, the maintenance craftsmen were reduced in number and reorganized into 'dedicated' teams, each with responsibility for a defined production area; they would no longer service the entire plant from a central workshop. The company called this 'the idea of ownership', with groups of maintenance workers being assigned direct responsibility for looking after defined areas of the plant. The stated intention was to reduce maintenance costs by 20 per cent and to reduce machine downtime, which in 1985 averaged 16 per cent – over 75 per cent caused by breakdowns. The maintenance workers' stewards, however, were concerned that the same system could be used as the basis for more effective monitoring – and hence control – of work through the closer identification of individual workers with specific tasks.

Second, the company proposed alterations to the system of allocating overtime among production workers. We will look at this in some detail, since it was this company's means of achieving production flexibility. Production work was undertaken by semiskilled men who did the loading, mixing and cooking, and women who did the packing. The full-time women had consistently been paid a flat rate for a 40-hour week anything between 5 and 10 per cent above the industry average. They worked virtually no overtime and were freely moved between packing lines as needed and, according to both the convenor and the personnel manager, had always done so without any attempt to impose restrictions or to claim reward for doing so. Packing work was all rated on the bottom two grades of the job-evaluated scale. The male process workers, all of whose work fell into the top four grades, had consistently received a basic rate around 10 per cent below the industry average, a rate that was low for the district. This was a clear company policy. They compensated through overtime, working between 1975 and 1984 anything from 12.5 to 20 hours a week. (The company calculated that out of around 4000 manual workers, about half of whom worked overtime, total overtime in 1985 was equivalent to over 600 full-time jobs.) This overtime could for some workers nearly double a base rate of around £100 per week in the mid-1980s. The problem that started to emerge as new work came on derived from the tradition that men would work overtime only on their own production lines. As production of some lines increased, while others did not, some operators started to resist management requests to transfer between lines, in case they were transferred to one with lower overtime opportunities. This was resolved by a detailed agreement on overtime distribution which committed the company to the broad principle of parity of gross earnings, in exchange for an agreement that overtime offered and not accepted (on any line) would forfeit further overtime opportunities. According to a personnel manager, this system now 'gives the company the flexibility to switch men from line to line, to close down lines when necessary, and to run two lines with one man' (fieldwork notes).

This system not only increased plant utilization, but it provided a partial answer to the second problem noted above – the need to increase shift-working. Some of this was achieved without taking on additional workers by increasing the amount of planned double-shift working on overtime. (Among the women packers the company introduced a part-time 'twilight' shift.) A joint 'overtime committee' was set up to monitor the new policy. A need for more continuous working had thus been achieved through an old device – overtime – rather than through relying exclusively on taking on new workers to handle the increased production. The benefit of en-hanced earnings to the process workers was thus to be partially set against the fact that relatively fewer new jobs were being created; a state of affairs acknowledged by a convenor whose hands were tied by members insisting on high overtime.[1]

These three devices, the 'de-manning' exercise, the creation of a new basis of maintenance, and the use of overtime to improve capacity utiliza-tion, were the major 'industrial relations' contributions to managerial plans to improve capacity utilization and to reduce costs. Their combined significance from company point of view was that they had effectively moved much of the work done by 2000 workers at the closed factory at very little extra cost, and with the creation of no new jobs in the surviving plant. The consequential intensification of work for the workforce in the latter plant was considerable, a fact acknowledged by stewards, and yet they hardly sought to contest it, let alone seek alternatives. The convenors' acknowledgement in interview that there might have been a union cam-paign for job creation in an area of high unemployment was overshadowed first by their insistence that the dominant (although fewer) men in the workforce were preoccupied overwhelmingly with overtime opportunities and, second, by their apparent acceptance of the managerial argument for a continuing need to hold down costs in the face of increasing competition from supermarket 'own-brand' products, a message frequently hammered home in consultative committee meetings, as well as in company videos shown to the workforce. This argument was accepted, despite the known fact that many of the 'own-brand' competitors were produced by the same company, some in a different factory in a different operating division, but some in the factory under study, on the same machines and by the same workers whose jobs were allegedly threatened by this competition.

Factory B
The company was a full-owned subsidiary of a large overseas company producing pharmaceuticals and general chemicals. The UK operation had five operating plants in this country, of which one was investigated during the research. At that time the factory employed around 650 process and 225 craft workers (electricians, fitters) and building manual workers. (The

company employed significantly larger numbers of staff and technical
workers, but they are not part of this study.) The manual workforce was
organized into one general and four craft unions.

The industry, although generally highly profitable, had been hit by a
number of changes in both national and international markets. Internationally, markets in the Third World had been lost as countries sought to
reduce their dependence on Western chemicals, drugs and fertilizers and
to build up their own manufacturing capacity. The first five years of the
1980s saw significant capacity reductions in most European chemicals
firms. In the UK producers were also hit by the high value of the pound in
a high-export sector and this put the UK near the top of the list for capacity
reduction and associated job loss.

In pharmaceutical chemicals, the market situation was slightly different
but the consequences not dissimilar. Established industry – government
relationships, with the National Health Service providing the dominant,
and apparently inexhaustible market, changed dramatically in 1984 and
1985 with the announcement of government plans to reduce profit targets
for the industry, and with the introduction of the 'limited list' of drugs that
could be prescribed through the NHS. This was aimed at saving £100
million per annum out of a total of £1.4 billion. The consequence, according to MacMillan and Turner (1985:14–18, 31) was not only to damage the
finances of companies affected but also to create 'uncertainty where once
there was stability . . . suspicion where once there was trust'. More recently some commentators have suggested that the savings achieved were
in fact nearer £60 million, and they have noted government plans to hand
back to the drug companies much that they had lost (see, e.g., *Guardian*,
10 August 1986). But at least for a period, the uncertainty engendered by
government action contributed to managerial proposals for cutbacks in
capacity and an overall lowering of activity.

This last point is worth stressing, since it demonstrates forcefully the
role that the state can play in defining the nature of 'private sector' product
markets. To pressures deriving from the forces of competition, therefore,
have to be added those of political decision, making managerial response
more complex and union reaction yet more problematic. In this case study,
the role of the state could be perceived at many levels. At national and
international level the regulations defining drug testing and safety procedures were vital, since the successful development of a new product –
capable of generating hundreds of millions of pounds profit – could be
frustrated or facilitated according to the strictness of national regulations.
As regulations between countries varied, and those within countries
changed from time to time, the giant multinationals that dominated this
industry could, and did, shift their centres of production. At lower levels,
the impact of the state could be perceived through, for example, differences in standard prescription and dose sizes, both of which affected
production technology. Similarly, regulations seeking to reduce the risk of

cross-contamination between products influenced both technological and labour relations developments, the latter, for example, being seen through restrictions on worker mobility between production lines in order to maintain sterile working conditions. All these factors impinged on the processes of labour relations and, for example, set limits on improvements in labour productivity.

The company studied has been no exception to the industry pattern. The parent company has responded to change through industrial policies of rationalizing production throughout its European operations, allocating markets and products to particular production centres, and either closing or selling off parts of the company that were producing at insufficiently high profit levels. Politically it, along with its competitors, has responded by lobbying to change UK government policy on prescriptive drugs, and at an international level by seeking to remove import barries. In several of these political campaigns it has sought to enlist the support of its trade unions and their sponsored Members of Parliament.

Reducing the Workforce: The Use of Consultation
In the plant studied the size of the process workforce had been steadily reduced from its 1979 level of around 1300 to the present 650; the craft and building workforce, however, had not been reduced as it was needed to help with the installation of new buildings and machinery. As with company *A* above, the major employer initiative had been to reduce capacity and output in line with falling demand and projections of a future that was, at best, uncertain. By contrast, however, this was achieved not by factory closure but by significant reductions within plants – including the shutting down of some lines and reductions on some others. This had been implemented through a policy of early retirement and voluntary redundancy, accelerated in both cases by terms seen as sufficiently generous to attract the necessary numbers. Indeed, in all cases, the applications for voluntary redundancy have exceeded management target figures. The unions' attempts to reduce the flood and to dissuade applicants have been in vain. In effect they have no policy on the issue and, as one steward remarked in interview, 'it is not part of a steward's job to tell people that . . . they should not take it [redundancy]' (fieldwork notes). A clear feeling emerged from interviewing stewards that the continuous stream of job losses, with the company continually looking for more, had led to a situation in which many workers felt that taking voluntary redundancy or early retirement, with cash in hand, was preferable to an uncertain future with the possibility of eventual compulsory redundancy. Stewards' occasionally-voiced unhappiness at the rate of job loss never materialized as a policy, those stewards who raised it admitting both that they had no membership support for it nor any power to enforce it.

The company did not negotiate redundancy procedures or compensation with the unions. Rather these emerged as company policy out of a lengthy and continuous process of consultation. The operation of the consultative process will be described in some detail, because it was far more than a mechanism through which views were sought by management. It was the prime mechanism through which management communicated their view of present realities and future prospects; it was, in short, the means through which the uncertainties engendered by product market change were communicated to the workforce, both directly and through its representatives. As such, it was crucial to an understanding of the processes of plant-level industrial relations.

The consultation process was one aspect of a major company initiative in the industrial relations area, dating back to the 1970s. It had been described in 1981 as being established to help achieve

> a more efficient company, able to prosper in the tough economic climate of today . . . in which [workers] will feel that they are involved and comitted . . . in which people understand what they are doing and why (company files).

But during the 1980s, consultation and communication had been seized upon and developed by management until, by the time of this study, both management and stewards viewed it as the central platform for their industrial relations activities. The hub of the process was the quarterly meeting of senior shop stewards, staff representatives (for the white-collar unions) and managers from all the factories in the company.[2] Subcommittees of this meeting were charged with discussing specific topics (e.g. redundancy procedures) and reporting back to the central committee. This process was supplemented by the printing and distribution of a monthly bulletin to all employees, intended to form the basis of shopfloor discussions led by supervisors. On some occasions these two processes overlapped – as when the senior managers announced a three-month wage freeze to the stewards, at the same time as section managers were informing the members in the factory 20 miles away. This episode clearly indicates the company's clear preference for relying on their own channels of communication to the workforce, rather than using the stewards.

The meetings of the central committee routinely revolved around the presentation of detailed performance statistics for the company as a whole and for each operating division. On occasions this led into announcements of proposed company policy to remedy perceived problems. Thus in one lengthy but not untypical meeting, the company produced a barrage of figures to demonstrate their poor financial performance and their bad labour productivity comparisons with their continental counterparts. This led into a clear statement that they would need to lose a further significant number (over 30 per cent) of jobs. Failure to do so, it was stated, might well lead the parent company to withdraw investment from the UK and locate it elsewhere. In reply to union protests that low productivity figures reflected antiquated and unreliable equipment, the personnel director retorted 'the company is bleeding to death. We need to share the bad news with our employees and start reconstructing from there' (fieldwork notes). Questions from stewards about the possibility of launching new marketing initiatives, or of pursuing political action to improve the company's position, were rejected by management as impractical and were not pursued by the unions.

It was clear from subsequent interviews with members of the shop stewards' committee that this announcement had not been unexpected. It represented the logical culmination of a succession of items of bad news. Few stewards questioned the logic of the company proposals; none suggested that the unions were capable of – or should even consider – putting forward alternatives. They knew that volunteers for redundancy would continue to come forward – several stated that they themselves were tempted. The unions were completely uncertain as to their response. The only certainties were provided by management – that the company faced serious problems that might eventually lead to complete closure, and that job losses provided the only possible means of salvation. The apparently insuperable inability of the unions to find sufficient common ground – both within and between factories – to enable them to come together to discuss any form of common response, exacerbated their problems. The relatively generous redundancy and retirement provisions eased both the departure of the workforce and the stewards' pangs of conscience at their inactivity.

Thus it was the consultative, rather than the negotiating, machinery that the company used to discuss and implement the more significant labour relations strands of their proposals. Wages and conditions bargaining continued at plant level on a fragmented basis. The company was prepared without much argument to concede pay increases broadly in line with inflation (just under 8 per cent in 1982, 5.5 per cent in 1984, 6 per cent in 1985), but their central concern throughout this period was workforce reduction. Thus although it could clearly be argued that in a formal sense the collective bargaining structures had remained unaffected by events, the importance of negotiation had been drastically reduced by a shift in context, to the extent that no real attempt was made by the unions to bring the new priorities – job loss and job protection – into the bargaining (as opposed to the consultation) arena. This example suggests that when some commentators argue that established bargaining structures can coexist with 'novel forms' of worker participation and consultation (see, e.g., Marchington and Armstrong, 1983) they may be missing an important point, namely that the major labour relations issues affecting the workforce have changed – they may no longer be wages and conditions – and decisions on these are now being taken outside the established negotiating channels and handled through consultation.

In this company, redundancy policy had never been a matter for

208 *Recontextualizing Shopfloor Relations*

negotiation. But before 1979, redundancy had not been a major issue. The big issues – from the members', unions' and company's point of view – were negotiated, especially wages. Since 1979 priorities had shifted, but the scope of bargaining had not been changed to accommodate the shift. The bargaining structures have not fundamentally changed, but the relative importance of the issues handled by consultation as opposed to negotiation has. In that sense, at least, it can be argued that consultation has displaced negotiation; unilateral managerial decision-taking had displaced joint regulation.[3]

It is important to understand – though difficult to demonstrate empirically – the extent to which the company's analysis of its problems, and its proposed solutions, were accepted by the unions, both in their collective dealings with management and in their internal discussions. Occasional grumblings, especially during wage negotiations, over managerial preference for investing in 'machinery rather than people' were never developed into anything resembling an alternative approach to the management of the company. For the workforce, therefore, the inevitability of the necessary remedy to the company's market problems – job loss – was reinforced by their representatives as well as their managers.

The absence of an articulated response by the unions cannot easily be explained in terms of a lack of resources (cf. Lane, 1982:11). The shop stewards' committee was, if anything, overwhelmed with information, not only from the company but from the industry committees of the unions on which the stewards were represented, as well as from attendance at briefing meetings organized by the Chemical Industries Association. The stewards had substantial local resources including two full-time convenors, few restrictions on several other senior stewards, and frequent opportunities to meet among themselves, and with stewards from other plants in the company, in work time and without loss of pay. They had access to offices, telephones, and printing facilities, producing several leaflets for distribution to the workforce. At the plant studied they made little use of full-time officials, but this was not the case in other plants.

Obviously much of the material received by the stewards was designed to support managerial conclusions. But it would be inaccurate to suggest that this bias prevented any kind of speculation as to other possibilities. Stewards strove to come up with remedies that obviated job loss, as indicated by some of their contributions to consultative meetings. But they appeared incapable of taking them beyond a first, tentative question. Their inability to deal adequately with wider issues of production was perhaps shown by their opinions of Friends of the Earth as mad people intent on destroying their jobs, a view they comfortingly shared with their managers.

It is well-nigh impossible to demonstrate the reasons for inaction. This case study suggests that it was not lack of material resources in the sense of time and information that inhibited the unions. Rather it could be traced to a political uncertainty as to how even to begin to formulate questions that would expose the inadequacy of managerial proposals, let alone come up with alternatives. The job losses and other changes were resented, and many stewards acknowledged that unions should be resisting, but eventual refuge was found in the 'apathy' of members and the inexorable logic of managerial analyses and remedies.

Factory C

As noted above, this factory is included in the study because, having gone through a period of severe job loss and dramatic production reorganization, it is now building up on the basis of new products and product organization, but in an environment significantly different from that it had known until the late 1970s. Prior to that date the factory, a small one in a large UK-based multinational, had for 40 years specialized in the production of large numbers of a small number of standard components produced for a few major customers predominantly in the motor vehicle sector. The work consisted of long production runs of uniform components for increasingly few customers. The process workers, predominantly women, worked on large banks of machines. There was very little need to move workers around, and output was maintained through the operation of a simple system of payment-by-results.

During the mid- to late-1970s the company's markets for these standard products collapsed through a combination of falling demand from established UK customers, and increasingly successful overseas penetration. In the five years up to 1980, around 300 manual jobs were lost, almost 50 per cent of the manual workforce, mostly through non-replacement as long-serving, elderly workers retired.

The plant might well have faced the prospect of complete collapse around this time had this been the only aspect of their activities. But during the 1970s the company had been experimenting with a range of new products using existing technologies. These appeared to be successful, and in the late 1970s the parent company decided to go for growth and expansion on a highly diversified product range. This move was so successful that by 1986 the company had on its books over 3000 customers in a wide range of industrial sectors, and management estimated that they were actively involved in producing components for around 600 of these at any one time. In recent years the company had started recruiting again in a small way and shopfloor numbers had climbed to around 350. For many of its products the company was now the dominant (if not the only) UK supplier, although they did face competition from overseas.

By the time of the research the changeover was virtually complete. As the convenor stated, 'we are no longer mass producers with a bank of girls turning out components. We are in the jobbing business. Customers . . . give an order and they want it now' (fieldwork notes). The broad contours of the labour relations implications had been outlined by the management at a works committee meeting in 1979.

We need to accept that the business is going from large volume production to batch production. . . . We need to accept changes in technology and manning requirements. The most noticeable changes will be the increasing versatility of the labour force and a ready acceptance of changed working patterns. In the short term a greater use will be made of the setter/operator category of employees and semiskilled operators are to be retrained for this (company files).

Moving to Specialist Production: Redesigning the Payment System
To a greater extent than in either of the previous cases, the management in company *C* explicitly saw labour relations reform as central to their new production objectives. In essence they sought two objectives: greater freedom to move operators between tasks to match production demands, and reducing the time and costs associated with frequent resettings of short batch jobs. From the early 1980s, the achievement of both was identified with the need to change the payment system.

As noted above, the traditional system had been to use individual piecework. But this suffered from the frequently-noted disadvantage of fostering resistance to being moved. Workers became familiar with particular jobs and had no wish to be moved to a new one where earnings might be at least temporarily reduced while they learned a new task. Management therefore developed an approach which looked to replacing the piecework system with one based on a guaranteed base rate plus a company-wide output related bonus that would eventually account for as much as 20 per cent of take-home pay. At the same time they were committed to the introduction of a single 9-grade job-evaluated pay structure covering all manual workers, to replace the simple groupings of skilled (mostly indirect) and semiskilled (direct) workers. This, it was hoped, would facilitate the blurring of distinctions between direct and indirect workers (operators doing some setting work, setters running machines), and encourage job flexibility. The suggestion was even made that there might be scope for supervisors to recommend the promotion of individual workers on the basis of their general preparedness to co-operate in the matters of flexibility and mobility.

Keen to ensure that the changes went through as smoothly as possible, and under no immediate pressure either from markets or from elsewhere in the group for rapid change, the company decided against trying to introduce all the reforms in one large package, preferring instead a gradual approach. Payment system reform was slowly introduced by recruiting all new workers on to a system of flat-rate pay plus a fixed lieu bonus related to historical piecework averages, and by transferring groups of existing pieceworkers onto this system as and when they agreed. They also increased the output-related bonus element of take-home pay over four years through the annual wage negotiations to its average level in 1986 of 7 per cent.

The introduction of a new grading structure could not be achieved on a similar incremental basis, however. Management therefore decided from the start to seek to win over the plant convenor to their proposals. This man, the only long-serving convenor encountered in these studies (he had held the post for 16 years), had retained his full-time status despite the reduction in workforce size. The personnel manager proposed through the works committee that he and the convenor should jointly survey the manual workforce, using a questionnaire designed to find out their views on the existing payment system and other matters. Fifty shopfloor workers were interviewed by these two men for at least an hour. The results of these interviews were that significant majorities of shopfloor workers were more concerned with job security than with basic pay, felt that piecework-based systems were associated with too wide a dispersion of earnings, and that one possible remedy for this might include a rational wages structure based around job evaluation and the elimination of piecework.

This exercise had the useful effect of confirming shopfloor support for managerial proposals and, at least as important, strengthening the convenor's hand in dealing with the resistance to the proposals that were anticipated, correctly, from certain sections of the workforce and the Joint Shop Stewards' Committee. This resistance led to the rejection in a workplace ballot of a pay and restructuring package in 1985. In another ballot, however, the membership made it clear that they were not prepared to take any form of industrial action against management proposals. The convenor, himself commited to the package, used this as evidence of the futility of attempting to resist, and in 1986, after a further campaign involving briefing meetings by managers and further pressure within the Joint Shop Stewards' Committee, a membership ballot narrowly (by a 55 per cent majority) approved a revised scheme in 1986.

The company appeared to have achieved what it wanted, at relatively little cost, although there was, by the time of the research, no evidence as to whether or not it would achieve all its stated objectives. But the example demonstrates forcefully the apparent usefulness to managers, under certain circumstances, of introducing change through 'traditional' collective bargaining. In this case the pivotal role of an influential, able and strong convenor, convinced of the need to move along the lines proposed by the company, was the crucial factor influencing the company's approach. The personnel manager clearly acknowledged this in interview. The convenor in turn justified his own activities (most of his work with management involved no other stewards) on the grounds both of the lack of ability among his fellow stewards and that there was no longer any chance of benefiting workers through 'traditional' methods. As he argued, 'the workforce are very worried about their jobs and any form of industrial action frightens them to death. . . . so no real bargaining with management can take place' (fieldwork notes). Three brief concluding observations can be made about this case. First, that, as in the other two factories, managerial definitions of problems and solutions were effectively unchallenged. Second, that, by contrast with the other two, labour relations considerations figured more strongly in management plans. Third, that at least in formal terms, the changes were introduced through long-standing procedures of collective bargaining.

Implications: The Limitations of Collective Bargaining

The studies confirm two broad contentions. First, that the three companies involved had, in their reactions to their particular market crises, implemented major changes to both the level and organization of their productive activities, that these had brought in their wake both job losses and work intensification, and that in only one case, the factory scheduled for closure in company A, had anything approaching a systematic union response been forthcoming. Second, that the changes had been achieved through stable plant-level structures of union representation, consultation and negotiation. A move against the workforce had been achieved without any subsequent hostility towards trade unions or union activists.

All the companies were more preoccupied with the exercises involving job loss and production organization than with bargaining over wage levels. Not only did the latter occupy little managerial time compared to 10 years earlier, but the tactics associated with such bargaining had changed. All companies were concerned to avoid any diversion from their reorganization plans by becoming involved in lengthy wage bargaining, even though the likelihood of industrial action over wages was universally seen as remote. Their offers were therefore pitched at a level intended to facilitate quick settlement rather than to pay the minimum possible. This usually meant an offer in line with, or slightly above, the current inflation rate. The bargaining that occurred between first and final offers through the established bargaining machinery was conceded by all to be predominantly ritualistic; none of the stewards interviewed argued that union action or argument had budged the company from offering what they wanted to pay. One direct problem for the unions that resulted from this was that the stewards derived no self-confidence from wage bargaining. Another was that they were unable to use wage bargaining as a means for bolstering their prestige in the eyes of the members.

Managerial attention focused upon reorganization of production. Several common themes emerged in the process of implementing the associated changes. First, all the companies were concerned to ensure that information was widely available to unions and members about the reasons for changes and the dire consequences of failing to achieve them. Second, that the underlying reasons were in all cases presented as lying outside the company's direct control, laid at the door of government, the behaviour of competitors, or 'the market'. Third, although workers were not to blame for the problems, they had to play a part – together with managers, shareholders etc. – in sharing the sacrifices necessary to improve company performance. In their case this meant acceptance of job loss and accept- ance of the need for increased flexibility and labour productivity. The only problem was that while rejection of such changes would, it was held, seal the company's fate, acceptance would not necessarily ensure success. Fourth, all the major elements in the companies' packages put forward to achieve these changes were treated, explicitly or implicitly, as non-nego- tiable items.

With the partial exception of company *B*, which had been in the process of revising its consultative machinery before its crisis broke, none of the companies had sought any changes in bargaining machinery or in union organization. Indeed, in the surviving factory in company *A*, management had gone to some lengths to maintain 100 per cent trade union organiz- ation, and in company *B* the time-off facilities for stewards had been greatly extended to enable them to attend all the meetings involved in the new system. The simplest assumption would appear to be that in all cases management felt itself to be under no internal pressure to disrupt a system that had not served them badly in the past; there was little point in change for its own sake. In the event, management in all three companies achieved their objectives through the established machinery, without resistance and, in several cases, with the additional authority of apparent – in some cases active – support from the unions. On the only occasion where management did encounter resistance, in the company *A* closure, the rhetoric of consultation was dropped to reveal the power relationships beneath.

It should be noted that all the union responses described were restricted to plant-level activity. This is not as a result of a methodological defect in the research; nothing else was going on in the trade unions. Stewards interviewed were simultaneously proud of their 'independence' and aware of its limitations, but invariably the finger of responsibility for the lack of company-wide activity was pointed at short-sighted stewards in other plants, or other trades. The crucial issue, of which many stewards were aware, was that their members were confronted with problems that could not be solved at plant level. While managers were able, in their presen- tation of information to the unions, to move from company- or industry- wide issues to plant-level policies, the unions, rooted in the plant, were unable to move from a plant-level understanding of members' interests to a company- or industry-wide policy. To do so required not only union structures at these levels, but the political and ideological resources capable of producing an alternative response at those levels. It is in that sense that a central argument of this volume, namely that restructuring presents a structural and ideological crisis for trade unions, can be per- ceived at plant level.

It is easy to see how, in the absence of a wider response, unions at plant level end up endorsing managerial proposals for change. For at that level, there often is nothing else that can be done. Calls for the maintenance of the *status quo*, so often a stock-in-trade of shopfloor trade unions, are unconvincing to both unions and members alike and, in the present economic circumstances, largely unenforceable. In the vacuum that remains, unions effectively follow managerial proposals, and devote their efforts to seeking to improve their terms of redundancy, or to retaining a degree of influence over redundancy selection. By thus publicly associating themselves with job loss, they further reduce the likelihood of coming up with alternative approaches.

The limitations of plant-based organization can be seen in two further ways. The 'alternative plan' at company *A* was a plan for survival at one plant based on a reduction in jobs and a deterioration in terms and conditions. Even if stewards in other plants had been sympathetic, there were obvious dangers in overt support, since it might have implied that they too were prepared to accept similar deterioration. No attempt was made to develop company-wide proposals that would have protected the threatened plant. Union weakness therefore lies not just in inadequate combine organization (see Lane, 1986), but in the failure to conceptualize a union response that might have a company-wide logic. The failure to develop the latter inhibits the creation of the former.

The limitations of such 'factory consciousness' for trade unions have been thoroughly noted (see Beynon, 1973). The difference between then and now is that in 1973 Beynon was also able to point to solid, albeit constrained, advances made by workers and unions on the basis of factory organization. More than a decade later, that is more difficult, since shopfloor unions are operating within a context of structural change where shopfloor organization is unable to deliver even the limited achievements of earlier periods. Employers are able to use arguments based on the logic of market forces to drive home the inevitability of plant-level adjustment, while the problems facing the unions are those of developing not only wider structures, but wider appreciation of the issues.[4]

It may be the case that although there are self-evident limitations on the ability of unions to respond in the present crisis, it remains tactically appropriate to do little, since retaining existing plant-based structures will enable a return to the *status quo* in collective bargaining terms when the changes have been made. This line of reasoning begs the central question of the present volume, namely that the changes to the organization of production precipitated by the present crisis are structural rather than cyclical in nature, and hence that future structures of employment will be qualititively shifted. We can perceive elements of this change in the case studies in this chapter. In company *A*, the emergence of a distinct 'own-brand' range of products and the complex structures of intra-company competition and production to which this gives rise, will persist as problems for the unions. In company *B* the international restructuring of markets and production location will persist. In company *C*, where the process has gone furthest, the changes are yet more clear. A production process based on steady demand for a range of standard products, with an appropriately-structured union organization, has been replaced by a 'jobbing factory' with many hundreds of clients demanding frequently-changing products. Production insecurity is thus likely to be a continuing feature of that factory. It is unlikely that union hopes of a return to a familiar *status quo* will be realized.

Notes

1. Members' demands for high overtime also undermined union claims for a reduction in the length of the working week.
2. There were no company-wide negotiations. The unions had very little contact between plants; there was no formal blue-collar/white-collar contact within plants, and quite a lot of craft/process friction. At these meetings the unions therefore had no agreed position, and on several occasions contradicted one another. There was no union forum for following-up the discussions that took place nor for formulating a 'union side' policy.
3. Similar arguments might be advanced in considering other aspects of labour relations which have gained in importance. For example, labour mobility, staffing levels and work allocation generally are often areas of managerial prerogative discussed, if at all, only through consultation. These issues are now sometimes seen as being of greater strategic significance to management than wages and conditions. Yet there is no evidence that they have been brought into the bargaining arena.
4. The law, in particular the Employment Acts of 1980 and 1982, further reinforce such factory-based isolation, restricting as they do the legal opportunity for many forms of extra-plant industrial action (see, e.g., Wedderburn, 1985: 43).

Bibliography

Batstone, E. 1984. *Working Order*. Oxford: Blackwell.
Batstone, E. and S. Gourlay 1986. *Unions, Unemployment and Innovation*. Oxford: Blackwell.
Beynon, H. 1973. *Working for Ford*. Harmondsworth: Penguin.
Daniel, W.W. 1987. *Workplace Industrial Relations and Technical Change*. London: Frances Pinter.
Hyman, R. 1989. 'Dualism and Division in Labour Strategies', *The Political Economy of Industrial Relations*. London: Macmillan.
Edmonds, J. 1984. 'The Decline of the Big Battalions', *Personnel Management*. March, 18–21.
Edwards, P.K. 1987. *Managing the Factory*. Oxford: Blackwell.
Guardian 10 August 1986.
Lane, T. 1982. 'The Unions: Caught on the Ebb Tide', *Marxism Today*. 26, 9, September, 6–13.

——. 1986. 'Economic Democracy: are the unions equipped?', *Industrial Relations Journal*. Winter, 321–8.

Levie, H., D. Gregory and N. Lorentzen Ed. 1984. *Fighting Closures*. Nottingham: Spokesman.

Marchington, M. and R. Armstrong 1983. 'Shop Steward Organisation and Joint Consultation', *Personnel Review*, Vol. 12, No. 1, 24–31.

MacMillan, K. and I. Turner 1985. 'Government-Industry Relations in the Pharmaceutical Industry'. Henley Management College, mimeo.

Millward, N. and M. Stevens 1986. *British Workplace Industrial Relations 1980–1984*. Aldershot: Gower.

Terry, M. 1986. 'How do we Know if Shop Stewards are Getting Weaker?', *British Journal of Industrial Relations*. Vol. 24 No. 2, July, 169–80.

Wedderburn, *Lord* 1985. 'The New Policies in Industrial Relations Law' in P. Fosh and C. Littler (Ed.) *Industrial Relations and the Law in the 1980s*. Aldershot: Gower.

Index

task recomposition, 6, 17
task specialization, 62, 72, 145
taxation, 69–70, 71, 81
Taylor, F.W., 138
Taylorism, 5, 145, 146, 155
teamworking, 146, 147–9, 150, 151, 155
technical control, 124
technical reorganization, 141–4
Technical and Supervisory Staffs, 73, 74, 170
technological change (textile industry), 94
  British context, 102–8
  changing nature of industry, 95–100
  conclusion, 115–18
  international patterns of restructuring, 100–2
  production restructuring, 108–15
technological unemployment, 13, 40, 56
technology, 4–5, 6, 16, 56
  group, 141, 143–4, 151
  robotics, 13, 124, 168
tendering, 65, 71
  fast tracking, 67, 72, 84
Textile Horizons, 102
textile industry, 15, 16, 17, 94
  British context, 102–8
  changing nature of, 95–6
  empirical patterns of change, 96–7
  forces for change, 99–100
  international patterns of restructuring, 100–2
  reconstruction of workplace production, 108–15
  theoretical perspectives, 97–9
TextileCo, 106–8, 110, 114–15, 117
Textiles, Clothing and Footwear Industries Committee, 109
Thermal Insulation Contractors' Association, 73
Third World, 3, 98, 102, 115, 117
Tootal, 105, 115
Toyne, B., 100
trade agreements, 73–5
trade associations, 34, 38, 41, 54
trade cycle, 41
Trade and Industry, Department of, 31, 35–6
Trade Union Research Unit, 113
trade unions
  automotive components industry, 123, 127, 150–5
  construction industry, 59, 61, 73–80, 86–8

fork lift truck industry, 162, 169–72, 176, 184–6
foundry industry, 22–4, 31–4, 37–40
  Lazard scheme, 44, 48–50, 55
  organization, 17–20
  response (case studies), 199–212
  restructuring and, 10–12, 13–14
  shopfloor, 191–6, 198, 212–15
  textile industry, 113–15, 117–18
  see also individual unions
training, 61, 80, 85, 111, 178
Transport and General Workers Union, 33, 73, 77, 86, 170
Treasury, 87
tripartism (fork lift truck industry), 172
tripartism (foundry industry), 53, 55–6
  operation of, 22, 24, 37–40
  origins, 31–7
TUC, 32, 39, 49, 55, 79, 109, 172
Turnbull, P.J., 136, 140, 149, 151, 154, 155
Turner, Dennis, 61, 62, 65
Turner, H.A., 113–14
Turner, I., 204

Uff, J., 72
Undy, R., 71, 76
unemployment, 18, 191, 193
  construction industry, 61, 68–9
  fork lift truck industry, 176, 185
  technological, 13, 40, 56
  textile industry, 96, 110, 113, 114, 117
Union of Construction and Allied Trades and Technicians, 68, 73, 76, 77, 86
Union Membership Agreement, 198
United States of America, 128, 129
upmarket restructuring, 94, 99, 116–17

vertical dependence, 16, 17, 123, 126, 136
vertical integration, 102, 104, 106, 107, 136, 173
Viyella International, 105

wages, 4–5, 12–13, 18
  automotive components industry, 123, 146, 150–4
  bonus, 74, 75, 76, 80, 171, 210
  collective bargaining, 212–15
  construction industry, 64–5, 74–6, 81, 82
  factory *C*, 210–12
  fork lift truck industry, 170–1, 188
  overtime, 75, 185–6, 202–3

226 *Index*

Index by Jackie McDermott